HEAVY METAL

HEAVY METAL

THE HARD DAYS AND NIGHTS OF THE SHIPYARD WORKERS WHO BUILD AMERICA'S SUPERCARRIERS

MICHAEL FABEY

wm

WILLIAM MORROW

An Imprint of HarperCollins*Publishers*

HarperCollins books may be purchased for educational, business, or sales promotional use. For information, please email the Special Markets Department at SPsales@harpercollins.com.

FIRST EDITION

Designed by Bonni Leon-Berman

Library of Congress Cataloging-in-Publication Data has been applied for.

ISBN 978-0-06-299625-1

22 23 24 25 26   LSC   10 9 8 7 6 5 4 3 2 1

To my wife, Barb; daughter, Megan; and son, Jason.
Thank you for being the font of my hope,
the source of inspiration, and the general
reason for getting out of bed every morning.

CONTENTS

AUTHOR'S NOTE

I remember the first time I saw the Newport News Shipbuilding waterfront, in the late 1980s. I was a cub reporter visiting the city for a job interview at the local paper, the *Daily Press*. I drove past the shipyard and pulled immediately into the first parking space I could find. In the dusky glow of the coming night, shadows and silhouettes of the cranes and a carrier island rose like specters from the waterfront. I had grown up in the big city, but I had never seen a cityscape like this before.

In the ensuing years, I drove past the yard anytime I was near the downtown waterfront, just to catch a glimpse of those cranes and carrier tops, hungering for information about the daily lives of those who worked behind those iron gates. I finally began to satisfy my appetite more than a decade later when I returned to that paper to cover that yard. I discovered a whole new world—some might say a forgotten way of life—where American labor molded American steel into American warships to ensure the American way of life.

Such a blithe summary, though, fails to capture the existence of those shipbuilders as they expend a daily dose of blood, sweat, and steel to build the mightiest warships ever designed and constructed. Those men and women put their bodies, limbs, and very lives on the line every time they walk through those gates. The carriers may garner the headlines through the years, but these men and women make that all possible in one of the most dangerous workplaces imaginable.

The more I came to know this clan of shipbuilders, the more I learned to respect them and what they laid on the line day after day. They took pride in what they did for America, but they also realized the consequences of an improper weld, poorly installed cable, or misaligned section of steel. American sailors could lose more than a sea battle; they could lose their lives. With that in mind, Newport News shipbuilders often risked *their* lives to do their jobs right and keep sailors safe in their floating carrier-city.

I wanted to tell the story of the waterfront and the men and women who worked there, hailing from all over Hampton Roads, Tidewater, and other areas of an obscure corner of southeastern Virginia, where a good day's work still counted as its own reward.

The tales of those unsung heroes deserve telling.

HEAVY METAL

PROLOGUE

The company picnic unfolded unlike any in Virginia before, with thousands of workers, their families, and their cars converging on a waterfront warren of small crisscrossing streets now patrolled and controlled by squads of local police. Nothing small happened at Newport News Shipbuilding and this fall day in 2011 would be no exception—the shipyard's first Family Day in about two decades to help celebrate its 125th year in existence and just a general acknowledgment of good days now and for the expected future.

The yard's siren blared, echoing across the city and beyond, but instead of tens of thousands of hard-hatted steelworkers pushing through the gates, an army of children led the initial charge, sporting smiles and shorts. Overhead, a spritz of drizzle dropped here and there from gray clouds, but no one cared. The rain refreshed—a late September day like this often heated into a sunbaked scorcher on the Tidewater banks of the James River. Most found the day to be perfect for county fair–like events.

That's exactly what awaited all those streaming, screaming kids inside those gates where their fathers, mothers, uncles, aunts, brothers, and other kin built the biggest warships of the American fleet. The youngsters lined up for an assortment of rides, a cornucopia of food, and a boatload of games that entertained the spirit and tested the imagination.

Seasoned shipbuilders recalled when the yard celebrated Trade Days inside the gates, and the workers really strutted their stuff, walking family members right up onto the carriers they were building or fixing. Forget about a rollercoaster; the kids had a real ride then—

something no county fair, no other company—could offer at a shindig like this.

"Years ago, when my kid was little, they let the kids ride the aircraft elevator," said Big Ed Elliott, who had started working as a full-time shipbuilder as welder about three decades earlier. The little ones would gather on the platform meant to carry the navy's planes and make the ride from the hangar to the flight deck and back, the coolest moments imaginable for a youngster.

"Those kids didn't give a dang-gone about the rest of the ship," Big Ed remembered. "They just wanted to ride the elevator up and down."

Still, workers loved the current Family Day. Being shipbuilders, though, they of course came up with some of their own little touches to improve things. Lee Murphy, another yard worker with more than three decades experience, noticed they were setting up a booth at which kids would try to land paper airplanes on a carrier cutout. He went home and pulled a six-foot-long toy carrier from his attic for the task. Of course, a shipbuilder would have such a thing.

Especially a shipbuilder working for Newport News Shipbuilding, the cradle of all modern carriers. Just a couple weeks before Family Day, the yard had mated the eight-hundred-ton steel stern to the hull of the newest carrier under construction at the shipyard, the *Gerald R. Ford*, navy ship number CVN 78. Shipbuilders on the waterfront were feeling pretty good about themselves, the ships they were building, and the current course of the US and its navy. But inside the US Navy, analysts remained apprehensive. They anxiously spied the actions of an emerging naval power in the Western Pacific, and they worried about the future of its fleet.

In mid-August 2011, about a month before the Newport News Shipbuilding yard blared its horn to kick off its Family Day celebration, a newly outfitted aircraft carrier blew its own horn half a world away, as it headed out to sea through a thick blanket of fog that enshrouded the port and shipyard of Dalian in Liaoning Province off the northeast coat of China.

The ship's eight boilers and four turbines revved up to two hundred thousand horsepower to turn the four shafts, churning up the sea in a milky gray and white that mixed with the fog in a ghostly wake. About a thousand feet long, the ship featured a new engine, radar, guns, and other systems that it would be testing that month on sea trials.

The ship now plowing its way through the Bohai Strait, east of the Yellow Sea, bore little resemblance, operationally, to the empty hull China had bought from the Ukraine decades before.

Called the *Varyag* (from the Russian word *Varyagi*, meaning Viking) when the ship was built for the Soviet Union in the 1980s, the incomplete carrier sat and rusted on the Ukraine waterfront after the Soviet collapse. A Chinese company bought the *Varyag* to keep it from being scrapped. The company said it planned to anchor the ship in Macau as a floating casino, a story few believed. China towed the ship to Dalian, to be finished and outfitted with modern operational equipment and systems.

China said it planned to use the ship mostly for training and research—a story that many naval experts found more believable, given the ship's older design, the need for logistical support, and the utter lack of experience in the People's Liberation Army Navy, or PLAN, in operating any carriers.

In Newport News, they laughed off the Chinese carrier, which was built to carry about two thousand people and relied on a "ski-jump" flight-deck design—aircraft revved to full power and blasted off a ramp to gain flight attitude, reducing the amount of fuel, bombs, or other weapons they could carry. US Navy carriers built in Virginia carried about three times as many crew members and used powerful catapults to fling planes off the ship, loaded with gas, bombs, and other payloads needed for combat. The way the American shipbuilding pros reckoned it, China was at least two decades behind the US. It would be years before PLAN could deploy this contraption.

Still, the overhauled and updated *Varyag*'s sea trials of August 2011 concerned quite a few in the US Navy. And in the US Pacific Command. And in the Pentagon. And in the military commands of just about every Asian nation, including long-time American allies and partners. With these sea trials, China was joining the very exclusive club of nations with operational aircraft carriers. Virginia shipbuilders might see the Chinese carrier as a joke; other naval strategists saw it as a warning. The US was the only country with a permanent carrier presence in the region—with one ship—and here was China testing one of its own while building another from scratch at that very moment, with plans for more.

A central question loomed: What did China plan to do with a carrier force? A Japanese defense study released at the time raised concerns about China's unwillingness to tell the world the extent of its military plans.

While Beijing remained officially mum on the subject, Japan, the US, and others could glean what the general thoughts might be among Chinese leadership from the popular and influential 2010 book *The China Dream: Great Power Thinking and Strategic Posture in the Post-American Era*, written by retired People's Liberation Army colonel Liu Mingfu.

A highly respected Chinese military strategist, the former PLA officer wrote, "In the 21st century China and the US will square off to become the champion among nations." He argued that "China and America are destined for fierce competition" and "the world is too important to give to the United States." Therefore, he said, "China must revive its martial spirit . . . a kind of spirit that dares to risk its life for the survival of a nation."

As an ideal, China Dream became the policy and mantra of President Xi Jinping, who showed his strong support for the ambitious carrier program.

In its own report assessing China's military, released in 2011, the Pentagon predicted, "During the next decade China is likely to fulfill its carrier ambitions."

The year 2011 would prove quite pivotal, too, for US carrier desires—and for the Newport News yard and workers who built those ships.

1

STEEL CITY

WINTER 2011

Interstate 664 tried to race past Newport News, high above the old downtown and its waterfront, but the traffic stalled as the on-ramps dumped cars onto the freeway and the off-ramp congestion jammed up the main transportation thoroughfare. It was from this vantage point that the breadth and scope of Newport News Shipbuilding could be truly appreciated. The shipyard monopolized the southwestern swath of the city limits, claiming the length of the James River banks, clutching the waterway with a series of piers and docks augmented by fort-like warehouses and construction sheds—a slice of the industrial Northeast erected on a Dixie waterfront, cranes' silhouettes rising from the waterfront like giant metal insects, including the skyscraper-ish, monstrous gantry known as Big Blue to honor its color, size, and power.

In Newport News, the city and the yard were indeed one. Once a seaport hamlet, the waterfront grew into the nation's naval shipbuilding juggernaut, simply called the yard throughout the southeastern region of Virginia known as Hampton Roads. Tens of thousands of steelworkers plied their trades there for Local 8888.

Cartographers etched the name Newport News on the map more

than a century ago, when sailors, dockworkers, and watermen roamed the streets watching bears dance and hunting women and whiskey in garishly painted wooden shacks that served as bordellos and bars— offering beer from vats you could wash a Great Dane in—along a downtown stretch called Hell's Half Acre and Bloodfield, near the newly built Chesapeake and Ohio rail terminal. On Sundays, evangelicals held religious services on Pier Seven, using rope coils and freight boxes for benches as the three-decked steamer *Virginia* belched on the river past oystermen and their raffish skipjack boats. Known as Smoky Joe, the steamer carried passengers destined for the Hotel Warwick, named for the county, which was one of Virginia's eight original shires.

Newport News Shipbuilding grew from the vision of Collis Potter Huntington, the American rail magnate who identified the spot as a terminus for his line after the Civil War. Huntington had first ridden up to the James River a century ago on horseback; while a teenager he had roamed the country as a peddler, selling and trading a variety of goods. The Connecticut Yankee surveyed the harbor of what became Hampton Roads and foresaw the future. Decades later, the peddler–turned–rail magnate gobbled up land where he expected his C&O line to end. At that time, it was mainly farmland marked by the earthworks to protect Confederate forces. Huntington needed a city to support his business plans and immediately set to work on the future layout as coal started shipping in 1882. He chartered his new shipyard— Chesapeake Dry Dock & Construction Company—four years later, the same year Newport News officially became a city. The new city's first mayor, Walter A. Post, later served as the yard's president.

Huntington was quite clear about why he wanted to invest so much in his new enterprise: "The roadstead, well known to all maritime circles, is large enough to float the ocean commerce of the world, easily approached in all winds and weather without pilot or tow; it is never troubled by ice and there is enough depth of water to float any ship that sails the seas and at the same time is so sheltered that vessels can lie there in perfect safety at all times of the year."

Exactly the kind of attributes that about a century later made it a perfect place to build the largest and most complex warship ever put to sea—the American nuclear-powered aircraft carrier, a one-hundred-thousand-ton sailing "city"—one with an airport for supersonic jets, powered by a pair of nuclear reactors the size of large vans, serving as a navy base with its own hospital, maintenance garages, fire department, and everything else needed for four thousand people to live, work, and make war, at sea, for months on end.

And on an unseasonably warm February day in 2011, with radio reports predicting highs in the 70s, shipbuilders marked an official milestone in building the US Navy's newest aircraft carrier—CVN 79. The ship would not be done and delivered to the navy for another decade, but given the recent carrier construction problems and the mood of Congress, the navy wanted to make sure there would be a CVN 79 to buy. The sooner the yard started building the new carrier, the better. The ship had no name yet, but today they would cut the first ceremonial steel plate—about the width of a steelworker's hand, as long as a pickup bed, and as blue as a summer's sky.

Shipbuilders cut the steel with a plasma burner, just one of the innovative everyday tools at Newport News Shipbuilding. The machine shoots a stream of plasma—the so-called fourth state of matter created by superheating a gas to temperatures greater than 21,000 degrees Fahrenheit—and then adding even greater energy with an electrical current that creates an electrified plasma arc at temperatures reaching about 40,000 degrees, which melts and cuts steel and blows away the molten metal. Equipment like this, in the hands of seasoned shipbuilders who had honed their craft over decades of waterfront work, made it possible to build the US carrier vessel nuclear, or CVN. Today the shipyard would be making the first cut on the seventy-ninth US Navy carrier, and yard shipbuilders knew they needed more modern innovations than plasma cutters to ensure that there would be a CVN 80, a CVN 81, and so on. The preceding carrier, CVN 78, the USS *Gerald R. Ford*, had proved so difficult, long, and expensive to build because of new designs, unproven technology, and Pentagon midcourse changes that if shipbuilders failed to build CVN 79 more cheaply and efficiently, the yard could be out of the carrier business, and they'd all be out of jobs. The future of the company, the city, and the region depended on building CVN 79 right.

Newport News shipbuilders wanted to prove they could do just that. With a newly configured digital waterfront that computerized almost every metal movement related to CVN 79, workers and managers bet they could build this carrier with fewer people in less time for less money than the previous ship.

CVN 79 was the first carrier completely designed in a computer so as to take advantage of digital engineering and manufacturing. For previous carriers, multiple paperwork packages were developed for each single portion of the ship, used to describe the assembly process but without much detail. The different steelworker trades and shifts had to physically hand off these packages between departments—leading to delays, misinterpretations, and production miscues. Easily accessible and shareable digital work packages changed all that. Thanks to computer modeling, experience with the *Ford*, and good old shipbuilder savvy, yard procurement managers could figure out not only how many thousands of valves or whatever the waterfront needed in total for the ship, but also how many it needed delivered in any given week.

To kick off negotiations with the navy to buy CVN 79, the yard promised to build it in fewer man-hours than CVN 78 took. A man-hour—essentially each hour each employee works to build a ship—was one of the most basic units of measurements in determining the cost to construct a carrier. Looking to slash CVN 79 costs all around, the navy expected to reduce the material cost of the ship up to 20 percent compared to *Ford*, to reduce the number of man-hours by up to a quarter, and to reduce the cost of government-furnished systems by up to 10 percent. It eventually took about forty-nine million man-hours to build the *Ford*, and the yard had committed to build CVN 79 with forty million, a seemingly impossible commitment. The plan had interlinked the fates of both ships—what happened on or with the *Ford* affected the CVN 79 program at a DNA level—design changes, system failures, and even the amount of work needed to fix CVN 78. To truly appreciate the difficulty of constructing CVN 79 and bringing it to life, one must account for *Ford* and its construction issues as well as its operational successes and miscues.

Despite all that rode on that first slab of steel going through the plasma cutter that February day, it was just one of the tens of thousands of steel plates stacked as high as houses throughout the yard fields, the concrete and asphalt storage areas served by yellow crane bars with magnet legs hanging below them, and in the fabrication shops, where steelworkers lined up metal sheets as big as billboards to be aligned by lasers and cut by robots the size of small sheds. For the finer work—the kind robots could never get quite right—steelworkers bent over plates, balancing their butts on buckets, spring-green hoods pulled over their heads, tools in hands, orange-and-yellow sparks flying over blue coveralls and steel-tipped brown or black boots, their sizzling metalwork mixing with the din of the whirring compressors, clanging metal, and other machinery. The shop air tasted like burnt toast. Smears of light filtered in from windows and electric illumination above, reflecting from the yellow crane bars moving overhead along blue rails the length of the cavernous shop. A white banner hung in the background with blue lettering: MADE IN THE U.S.A.

The metal for the first CVN 79 steel cut in February 2011 had been in the yard for a while. The navy had started to order the carrier's steel and nuclear-power-plant parts from around the country about four years before the cutting began, with veins of railroad tracks and arteries of roads packed with shimmering sheets of steel from the mountain bowels of Pennsylvania and Indiana feeding in and out of the massive shipyard. At the same time, a small flotilla of steel-laden barges had made its way from the intercoastal through the Chesapeake Bay and to the James River yard depots. Altogether, a half-billion dollars' worth of special steel, forty-seven thousand tons of it of different widths and strengths, had arrived in the yard—mainly by truck, pancaked rectangular slabs up to four inches thick, to eventually be cut, molded, and assembled by waterfront women and men into an aircraft carrier.

The warship-city was itself being built by a city within a city. Newport News Shipbuilding lorded over its own little sovereign industrial hamlet of more than twenty thousand steelworkers of as many races, religions, and other diverse markers as one would find in any US city of that size, a city whose population could rise as high as thirty thousand and dip down to seventeen thousand, depending on US Navy contracts.

Many of those shipbuilders commuted well over an hour each day each way, including those driving down Route 17—known as it sliced north through Virginia as Tidewater Trail—alongside some of the very trucks stacked high with carrier steel, through the old established English-sounding counties like Middlesex, Gloucester, and York and the rest of the Virginia Middle Peninsula, where bald eagles often soared overhead and deer breakfasted near the side of the road in the predawn light with silhouettes of tractors, farm barns, and far-off tree ridges etching a relief on the shadows of the horizon. Even this late in February, as the yard prepared that metal slab for the first CVN 79 cut, some single-farm spreads sported Christmas lights and other decorations. The lights still twinkled as shipbuilders drove south toward Newport News, alongside well-worked pickups full of crab traps and tired watermen.

Most tuned in one of the many country or gospel stations in this

Bible Belt loop of Tidewater. Some listened to sports-talk radio about the Nationals' chances, or the lack of them. For those who cared, the morning news was not good. Libya remained a mess, following the UN Security Council sanctions on Colonel Muammar al-Qaddafi. Congress still fumed over President Obama's almost $4 trillion budget proposal; Republican lawmakers wanted major cuts—not good news for a yard that built the country's biggest and most expensive warships.

Ed Elliott III, known in the yard as Little Ed—but more commonly called Wingnut, because his ears stuck out like a couple of open car doors—liked to listen to all kinds of music, but particularly recorded live classic rock from the eighties, like the *Eagles Live* CD from 1980. On the day they prepared to cut the first CVN 79 steel, as on other days, he made the morning trek from Gloucester County in his 2004 Pontiac Sunfire. Little Ed now worked as a foreman for CVN 78 USS *Gerald R. Ford*. His father—Big Ed—worked as lead general foreman for *Ford* and had been a full-time shipbuilder since 1981. While Little Ed most certainly inherited his love for shipbuilding from his dad, as for Wingnut's ears, the elder Elliott would tell folks, "He got those from his mother." Many times, when Little Ed arrived at work, he'd find handfuls of wingnuts left anonymously on or in his desk.

Both father and son resembled pioneers from a faded black-and-white photograph, rigged for hard work by sturdy frames, square shoulders, and that certain way many Gloucester farmers and watermen had of carrying themselves, with no spare movements, no unnecessary lists. Big Ed sported a robust mustache, and Little Ed had sprouted a full but neat beard. Little Ed was a born wrench turner who now oversaw, built, and troubleshot piping for *Ford* carrier systems. Every time he walked through the yard gates, his eyes still sparkled with that sense of awe and pride he had felt that very first time.

"Why don't you try the Apprentice School instead of going somewhere to college?" Big Ed asked his boy as the lad prepared to finish high school. The yard had established its apprentice system in 1911 and by World War II had become the only institution of its kind with a junior college rating.

For Little Ed, the yard had been like a home away from home. He spent his younger years climbing half-built carriers—blown away by their size—and then attended their christenings. He never forgot the drive-through tour of the yard in a small company van, thinking, *I didn't know it was* this *big!* At the Apprentice School, they showed him everything—new carrier construction, carrier overhauls, the works.

Big Ed started in the yard as a welder and quit after two weeks. He came back in 1981 as an outside machinist and also attended the Apprentice School, graduating in 1985. He became an acting foreman as a third-class mechanic before being promoted to foreman—all the while making those daily long commutes with plenty of other shipbuilders. Sometimes they'd try to throw packages of Twinkies into each other's truck as they were going over the York River on the George P. Coleman Memorial Bridge.

Some yard commuters would drive through Gloucester to Mathews County. If you wanted to get to Newport News from Mathews, you had to go through Gloucester—there were few straight-line passages in Tidewater. The various bays, rivers, and other coastal features presented natural obstacles to most straight-line commutes. Roads often just dead-ended into some body or ribbon of water, one of the reasons that Newport News ran long and narrow along the James, in a shape like an aircraft carrier. Adjacent to Gloucester, Mathews remained even more isolated by the waterways and governed by the tidal ways. A "come-here"—anyone not born in Hampton Roads—stood out even more in Mathews.

Born in Mathews more than a half century ago, Lee Murphy never imagined living in any other place, nor working anywhere other than in the shipyard. By the time the yard was set to cut steel for CVN 79 that February, Murphy had worked his way up to superintendent for CVN 78 steel work.

Murphy chose to be a shipbuilder at age eighteen in 1976, a month after graduating high school, starting in the X-11 trade—the steel side of the house, as yard workers called it. He followed in the footsteps of his older brother, who then worked as an O-43 millwright.

Lee's focused now on erecting the individual units, the steel building blocks, that the waterfront arranged and combined and welded to create an aircraft carrier. As he'd explain it to outsiders, he did "all the erecting." He came just as the yard was expanding, building up the new North Yard—the womb of future aircraft carriers.

While more than an hour away from his small Mathews hamlet, the yard loomed large for Lee, as it did for many in the county or even farther away because of the pay, work availability, and at least some kind of job security. He possessed a wiry, flexible body typical of many longtime yard steelworkers. He squeezed into any space—a desirable trait for his metalworks trade—and he lasted in those cramped conditions for hours. He used his head as much as any college grad to work out the intricate geometry of the pieces in his units. He rose to be a foreman in 1983. The job provided what he valued above anything—the ability to raise a family in a good home.

His son Jason, though, harbored no long-term ambitions as a youngster to follow the family's shipyard legacy, despite Lee's attempts to make the waterfront part of family life. When Jason got a toy aircraft carrier as a toddler one Christmas, Lee stood over the ship like a huge human crane, hanging down his arms to cradle his ships in his claw-mimicking hands, bellowing out loud siren noises—*WEEEERH! WEEERH!*—sounding like the yard's giant crane during a carrier lift. When father and son later were building a treehouse in their family oak—a tree so big that it took six full-grown adults holding hands to encircle it—they did so in a *modular* fashion, that is by assembling certain sections first and then mating them together, the same way Lee had erected his carrier steel units.

Jason worked summer breaks in the yard, but when he graduated from Christopher Newport University, just down Warwick Boulevard from the yard's gates, he became a schoolteacher in Gloucester, following in his mother's wake instead. He was a Mathews man, though, and the ways of the water have a way of calling those men back. He enrolled a few years later in the yard's Apprentice School and entered the yard, not in his dad's metal trade, but as a designer, the job he had that winter day in 2011 when the yard officially started CVN 79.

Murphys, Elliotts, designers, foremen—shipbuilders—all converged on Route 17 in February 2011, with trucks carrying steel, farm tractors, and other Middle Peninsula commercial vehicles on the Coleman Bridge—drivers praying there'd be no bridge openings—down to Route 105, or Fort Eustis Road, toward the storied army base to hop on to Interstate 64 to bypass all the northern Newport News traffic for the seven-mile stretch to the off-ramp for the highway leading downtown. There, steelworkers and steel each made their way into the yard, the trucks lining up at their various gates to offload and the workers battling for parking spaces along the streets and in the lots before heading to their own respective sidewalk gates, an army of workers packing the parking lots, sidewalks, and street crosswalks.

Whether workers drove into the yard from the road or walked in through the gate, the effect was the same—an all-out assault on the central nervous system. You kept your head on a swivel, no matter what your job. No amount of ear protection stifled the whistles, screeches, bangs, sirens, whirring machinery, and clanging metal. They had entered a battle zone, with no quarter granted for any mistake. They enjoyed no downtimes, no leisurely lunches, and no dawdling as thousands of workers raced throughout the yard on bike, foot, or one of the waterfront fleet of construction vehicles.

Unloaded from the northern trucks with loud clangs, steel meant for the first CVN 79 carrier sections—mostly in ten-by-thirty-foot slabs—was stockpiled around the waterfront and moved by forklifts and cranes into the fabrication shops, some the size of city sports arenas with bats flitting about, to be cut into the carrier's structural plates and shapes for the assemblies.

Outside the steel-cutting shop, workers and cranes stacked up steel plates in the fields and racks near the bays, a stadium-size Home Depot of steel, until electromagnetic cranes carried the plates to robotic cutting machines, with all the information on cut size, location, and shape fed into the computers controlling the robots' plasma scalpels. Beeps and sirens echoed through the hangar-like caverns as orange-overalled workers polished off some pieces before sending them over to carrier staging areas. Robotic cutters might be high-tech, but the industrial beasts still couldn't quite handle the finer work needed for navy carriers. Bending over steel beams, steelworkers burned out end cuts, finishing off bevels with hand torches, creating fireworks of sparks that showered their orange suits. The room reeked of burnt metal and smelled faintly like a car repair garage. At another bay, the yard's enormous metal presses and additional plasma machines shaped metal plates into carrier foundations.

Over in the shop's "big bay" stood a unique plasma-burning machine. It looked a bit like a pickup engine and transmission hung up by the tail end. It was used to cut heavy, thick steel, like the first cut that February 25, 2011, on CVN 79, made by Wayne Kania, a shipyard machine specialist. He dialed up the right data points into his computer, and the machine fired a flaming mix of oxygen and propane onto the 2.5-inch steel slab as navy, shipyard, and local politicians looked on, cutting two bevels with an electrified plasma arc to weld the plate to another to form part of CVN 79's hull. It was all over within a matter of minutes with little fanfare, but from this small beginning would arise an American aircraft carrier.

At the nearby panel-line shop over the next months after the cut, robotic joining machines revved up to mate and weld giant square plate panels—some, like those destined for the carrier flight deck, as wide as sixty feet. Some shaped or curved metal plates could not be moved through the automated joiners and other equipment. For pieces like that, steelworkers manually welded the pieces, held in place by overhead gantry cranes. The cranes laid the panels across the floor like pieces of a huge jigsaw puzzle and hooded steelworkers knelt in position, welding beams in place, creating a small molten stream in their wake.

"There was nothing like a Southern weld," Newport News welders boasted. These shipyard steelworkers became masters of electronic-arc welding, creating a circuit between the steel and a wire electrode to make the weld. The welder held the electric wire just a few inches from the steel, sparking a high-temperature arc, which cooked the edges of the steel and the electrode tip enough to fuse them together.

Thanks to computers, on CVN 79, steelworkers now did things with panels on the floor in the panel shop that they could not have done on previous carrier plates. They shot studs, hundreds of thousands of them—in a much easier position, leaning over the steel plates and shooting down. Previously, steelworkers climbed ladders to shoot studs in overhead spaces already built out on the ship. That change, though small—led to major savings in time and money, the kind of change that could save the carrier-building program at the yard, which was then in jeopardy.

Outside the panel shop, on the waterfront Final Assembly Platen, assembled plates and panels—the makings of a carrier's inner bottom unit, bilges, and lower side shells—stood erect and ready for dock assembly. There they were in the province of Lee Murphy, who, in his mind's eye, arranged those massive steel components into the compartments, units, and basic building blocks of the ship's hull. "It's just a matter of simple geometry," he explained to an outsider. He just had to maintain his focus from the very first ship unit until the ship launched.

Steelworkers worried a bit about the CVN 79 steel because, to save weight on the *Ford*, the navy and the yard had opted to use a lighter-grade steel, but it proved too thin for the early stages of construction for that ship. Shipbuilders remedied the thin-steel issue by building extra framing to support these massive assemblies, adding time, manpower, and cost. The yard—and the US Navy—could ill afford any extra cost. The *Ford*'s price tag had already hit more than $12 billion and, as a result, a *real* push had built in Washington, the Pentagon, and Newport News to drive costs down for CVN 79. The new carrier had larger and more heavily outfitted units and superlifts—massive ship sections joined for the biggest and heaviest yard crane lifts—than the *Ford*, to save costs, so workers welcomed the sight of thicker steel so early in the building of the carrier. CVN 79 needed about twenty-five thousand full-size steel plates, ranging in thickness from three-sixteenths of an inch to six inches.

Comparisons continued constantly between CVN 79 and CVN 78. That previous carrier, the *Ford*, had been the lead ship of a new class of carriers—the Ford class, which was designed to be much more technologically advanced than the existing Nimitz class of carriers developed in the mid-twentieth century. Due in part to the technology, the *Ford* was problematic to build. CVN 79 was the second Ford-class ship, and yard and navy promised cost savings and other benefits for the construction of the new ship beyond the normal lessons learned for a second carrier-class ship. Fulfilling those promises, the yard hoped, would erase the ill will created by the *Ford*, save future carrier contracts, and ensure work for the yard and jobs for shipbuilders.

Building the new carrier at the lowest cost but highest quality dominated the thoughts of Mike Petters, the Northrop Grumman executive and president of the shipyard when the yard cut the first steel for CVN 79. Petters knew how navy ships operated, from firsthand experience, honing his shipbuilding smarts for more than twenty years.

He was born on Christmas Day 1959, a son of orange and cattle growers in the rural community of St. Joseph in northeastern Pasco

County, Florida. The eldest of six, he learned early to serve others—a mind-set drummed into him by his father and reinforced by the Jesuits who schooled him. His dad, the son of a German emigrant, repeated one refrain to his kids: "You really are lucky to be born in this country, and before you go off and do whatever you're going to do, you're going to serve."

For Mike Petters's father, who rose through the National Guard ranks, that meant military service. The future shipbuilding executive picked the navy—he saw nuclear power as the future of the country, and the navy was cutting the edge in that field. He graduated Annapolis and served as the reactor-control officer aboard the strategic nuclear-missile-armed submarine USS *George Bancroft* (SSBN 643). Like any other officer aboard a nuclear-powered submarine at that time, he had to survive interviews by the infamous Rear Admiral Hyman Rickover, who threw Petters out of the office during their first meeting, giving the young officer a lesson on how to learn lessons—something that would serve Petters well through the rest of his life and career. After that first failed meeting, he learned to hold his ground with frankness and honesty. Petters later told one of his Newport News shipbuilders, "Once you've been chewed out by Rear Admiral Rickover, anything else is pretty lightweight."

It was the desire to be home with his wife and children, not the ire of Rickover, that drove Petters out of the navy. Deployed at sea when his eldest daughter was born, he wanted to be at home for future family milestones. During his last year in the navy, he served as the refueling officer at the Charleston, South Carolina, shipyard—his first taste of waterfront work. Like anyone in the navy, he knew of Newport News Shipbuilding. His wife grew up in Franklin—about an hour away from the yard. He also liked the yard's proximity to William and Mary, which offered one of the top MBA programs in the country, a degree he planned to pursue. Petters joined Newport News Shipbuilding in 1987 in Los Angeles–class attack-submarine construction.

Now in 2011, about two-and-a-half decades later, as the CEO and president of the new Northrop Grumman spinoff, Huntington Ingalls Industries, anchored by Newport News Shipbuilding, Petters tapped all his shipbuilding savvy and corporate acumen to build a carrier at a price that would satisfy the navy, a profit that would mollify investors, and a quality that would meet Tidewater shipbuilders' standards.

When Petters started at Newport News Shipbuilding, one of his mentors had hung a sign on his bulletin board that read, WHY IS THERE NEVER ENOUGH TIME TO DO IT RIGHT, BUT THERE'S ALWAYS ENOUGH TIME TO DO IT OVER? He had one mind-set from that moment on: "Let's get it right the first time." With CVN 79, there would be no do-over time. With all the questions, issues, and angst building over the *Ford*, the yard could ill afford *not* to get it all right the first time. Because any serious mistakes could lead to the end of big-carrier construction.

Still, Petters knew that the best-laid plans of shipbuilders often go awry. Nothing really comes together in exactly the right sequence at exactly the right time at exactly the right cost. In the best of times, he needed someone in charge to be creative. For CVN 79, that person would have to be *extremely* creative and also agile, ready to develop workarounds and re-engineer on the fly. Petters knew and understood ship production at the waterfront level. His online bios showed him in a suit and tie, but he struck folks as being more at home in a hard hat and safety glasses. He identified with those who worked along the banks of the James.

"They are able to do something that most of us don't get a chance to, and that is to take raw material and somehow with their hands transform it into something that is greater than themselves," Mike Petters once told Tom Clancy's writing partner, John Gresham.

"They do that with their hands. They just didn't wake up and say, 'I can go do this.' They had to learn how to do the shipbuilding trade. They've not only had to work with their hands, but also, they have a lot of knowledge and intelligence in their head in this. We have craftsmen who can run their fingers across a plate and tell you whether it's flat or not. Every single thing they do, they put their hearts into it." He believed in those workers' abilities to get to a production run for Ford-class carriers, starting with CVN 79.

When he first started in the sub-construction program at the yard, he considered subs the most complex things to build. Then, when he ran the carrier program, he became convinced that the carriers were the most complex. Shipbuilders managed the construction, assembly, and proper placement of thousands of compartments—and the thousands of people doing that work on an hourly basis. After all, the carrier nuclear propulsion plant alone equaled roughly five submarine propulsion plants. Thousands of welders, painters, and other trades workers doing every kind of construction work built, erected, and pieced together the three thousand compartments that formed a carrier. You had to make sure everyone worked in the same direction, though not all at the same time in the same place.

Petters lay there at night in his single-story brick home on the James River, just down the street from Hilton Village, and tried to sleep. The yard and the US government developed the little historical enclave of largely brick homes about three miles north of the shipyard to help provide housing for yard workers about a century ago. His mind kept wandering back to CVN 79. He had to get it all done right the first time. Do that and they could start getting back to building a carrier every four years, or at least every five. Defense Secretary Robert Gates wanted to build them every five years. It was closer to seven now. Shave off those two years and you had cheaper ships and fewer lawmakers, analysts, and so-called experts calling for the end of the ships.

To get to five years—or better yet, four—Newport News Shipbuilding first needed to get and keep CVN 79 on the right track from the get-go. The yard had just the man to do it.

Like other shipbuilders, Mike Butler made the most of his own long commute. Actually, he enjoyed his daily trek to Newport News

Shipbuilding from the southern Virginia border. From a family that originated in North Carolina and Georgia, Mike lived as far south as he could in Virginia without violating the state line. He used the extra drive time to mentally prepare for the day ahead or to unwind before he got home in the evening. He had been making the commute for over four decades, in either a pickup or a Jeep. Mileage be danged—his vehicles worked for a living. As he did.

Horizon-stretching flat farmland dominated much of this corner of Virginia, scenes that evoked a sense of Heartland, America. A farm boy by heritage, Butler felt at home motoring past the fields. He rolled down the window a tad and, a Southern Baptist to his core, he found morning Christian messages on the radio to soothe his mind. Fighter jets from nearby bases circled overhead as he made his way toward the yard on North 10 to Route 17, then over James River Bridge, the wind whipping up whitecaps.

Crossing the bridge, Butler thought of the job the yard had just offered him that spring of 2012: supervising overall construction for CVN 79, which the navy officially named the *John F. Kennedy* on May 29, 2011, about three months after the first steel cut, to coincide with the former president's ninety-fourth birthday. It marked the first time that a modern carrier would be named twice after the same person.

Butler, a former yard ship-repair supervisor, took charge of getting it right the first time. When he accepted the assignment to lead CVN 79, senior leadership emphasized simple expectations: *Find every possible way to do it cheaper and faster. Of course, do it right and with quality.* They wanted to develop a vision beyond the way the yard normally built aircraft carriers.

The entire future of the US Navy's carrier-building plan—and thus the future of Newport News Shipbuilding and the thousands of Local 8888 steelworkers there—depended on what happened in Dry Dock 12 with CVN 79, which would represent a reboot of the Ford-class construction program. Hit those goals, and new carrier contracts would be guaranteed—there was even talk of a two-carrier contract, something that had not been done since the 1980s. Miss the goals and . . . many other Navy program executive officers circled carrier funding, like sharks.

In the yard, the CVN 79 team reconsidered *every* step of carrier construction. *Kennedy* shipbuilders roamed the yard with a new set of eyes, scrutinizing the fabrication shops and asking questions: "How do we cut steel faster and better? How do we build bigger superlifts?" They scrutinized the platen by the side of Dry Dock 12 in the North Yard—the biggest construction dry dock of its type in the hemisphere: you could put the Empire State Building in there with room to spare. He asked, "How do we install wiring, cabling, and other vessel outfitting cheaper, faster, and earlier?"

To do this right the first time, Butler needed all the shipbuilders on board, buying into this whole new way of looking at carrier-building and realizing that the yard was serious about doing it differently. He made it clear he welcomed any innovative idea on how to build CVN 79. While he patrolled the dry dock one day, for example, one of the riggers approached him about getting a special forklift with a sixty-foot extendable boom. "I bet we can eliminate half the crane lifts on the platen," the rigger said. While those forklifts did not come cheap, Butler thought, "If I don't go through with this, these guys will never give me any more ideas." He got the forklifts, which the yard wound up using all the time.

Butler and shipbuilders exchanged waves, high-fives, and shouted greetings as they passed each other on the waterfront. "Hey, Mister Butler! Howya doing today, Mister Butler?" He came from the same kind of stock that many of them did. Butler's father started at the yard as an insulator, wrapping valves in insulation, minute after minute, hour after hour, day after day, week after week, year after year. A dull, dirty, dingy job. He impressed his supervisors with his work-hard-do-it-right performance, the desire to simply do a good job. When Mike

Butler sought work in the yard, his dad encouraged him. Like his father, Mike sought simply to be a good shipbuilder.

To him, being a good shipbuilder meant believing in and becoming a part of something much bigger than yourself. And that about described the team of workers Mike Butler and the yard assembled to build the *Kennedy*. No loafers or clock-watchers—they cared about their jobs, their work, their ships, and their country. "They're patriots," Mike Butler would tell outsiders. But he really saw them as moms and dads, churchgoers, Little League coaches and, of course, neighbors, in and out of the yard. No one else could do what they could—build a nuclear-powered aircraft carrier. Butler roamed the yard, his lean frame making quick sure steps through a minefield of metal along the waterfront, imagining with his team what most could not—these beams, sheets, metal pieces, and equipment now filling up Steel City being manhandled, molded, and manipulated faster and cheaper into a hundred-thousand-ton warship, the largest and most sophisticated vessel ever put to sea. Butler regarded his work to be a cool but humbling job. With his thousands of veterans of steel, as waterfront workers called themselves, Butler would spend the better part of a decade—first in Steel City and then in Dry Dock 12, creating a carrier and getting it right, the first time.

To ready the waterfront specifically for the Ford-class ships, Newport News Shipbuilding had done a bit of a makeover for Dry Dock 12. The *Kennedy* would be the first real carrier to reap the full benefits of a new elevator and additional ramps connecting the dock wall to the ship to make it easier to move shipbuilders, equipment, and material in and out of the dock and the vessel. The yard had also just erected two additional blast-and-coat buildings, tripling the amount of space available doing that work on assembly units. The yard replaced welding equipment and built new covered facilities to protect their work from the weather.

In January 2013, riggers the size of NFL linemen forged by frozen winters and boiling summers on the James scampered down the metal scaffolding stairs five stories deep into an open waterfront quarry of sorts to prep the dry dock to start the *Kennedy* construction, bracing themselves against winter blasts that could cut a man in half. This cradle of carriers stretched out more than 2,200 feet from the shoreline into the river channel. Anchored by a foundation of thick concrete, the whole structure was built on a landfill and secured by pilings pounded deep into the silt and bedrock several hundred feet below. At the end of the dock, a removable steel box caisson kept out the river water until the yard needed it.

The James River is a mental bookmark for anyone who had even a passing interest in the history of the United States. English settlement of North America took root with the establishment of Jamestown, on the James, after landfall by the *Godspeed*, the *Discovery*, and the *Susan Constant* at what was then called Point Hope, under the command of Christopher Newport, from whom—according to the most commonly accepted legend—Newport News got its name. It was where folks from the Old World and New World traded news of the known world. It was where a nation was built—and where the nation's biggest warships would come to be built.

The building of the *Kennedy* in Dry Dock 12, though, began just as it did for any other large-deck carrier for more than half a century. Riggers erected the 450 pillar-like foundations, which stood like concrete anthills or cairns as tall as the steelworkers, on the bottom of the dock, topped with a special type of red oak meant to slowly be crushed under the warship in a precisely arranged sequence to spread the weight and support the carrier as it was being assembled section by massive section.

2

OUTCLASSED

SPRING 2014

Walking across the flight deck past the island on the new carrier *Ford* in the spring of 2014, just as the yard truly ramped up to build the *Kennedy*, Rear Admiral Michael Manazir eyed CVN 78's tower, which housed the bridge, the flight observation deck (aka Vulture's Row), and its new sensor suites. Lacking the rotating radars of previous ships, the *Ford* instead had the flat panels of a more powerful and efficient radar that could scan the skies on two different bands, while also giving the tower and the carrier a sleeker look. The island sat farther astern than its midship location on the *Nimitz* and carriers before it; the new arrangement provided more acreage for flight operations. The tower looked like a castle.

Manazir—call sign Nasty, thanks to a mispronunciation of his last name back in his academy days in the 1980s—zippered his brown leather navy flight jacket against the unseasonably chilly spring breeze cutting across the flight deck from the James and took quick, careful steps, taking care to avoid the thicket of wiring, hoses, and other construction equipment on the deck and in the island, including the brand-new, in-deck fueling stations, one of the many new design mods to enable the *Ford* to launch and recover aircraft with greater speed.

Nasty simply loved this new class of carrier. The navy's air-warfare director made the three-hour car trip through Virginia between the Pentagon and Newport News as often as he could, for in-person updates on the *Ford* and the *Kennedy*. He needed all the ammunition he could get for his Beltway battles to save the carrier program, under the torturous grilling of the likes of Arizona senator John S. McCain, the powerful chairman of the Senate Armed Services Committee, or SASC, as it was commonly known in DC.

A number of Pentagon officials and lawmakers wondered why the US needed to keep building and deploying the nuclear-powered behemoths. After the collapse of the Berlin Wall and dissolution of the Soviet Union, the US Navy emerged as the uncontested king of the seas. Why the need for an expensive aircraft carrier? Many powerful members of the House and the Senate saw the current Nimitz-class carriers designed during the Vietnam era as floating anachronisms and money sponges.

A long line of polished, dedicated, and decorated admirals knew what it felt like to be dressed down by Senator McCain during an SASC hearing, his finger pointing, his voice rising, his ire building, like a prosecutor, interrogating, accusing, and damning. And *no one*—no matter what rank or title—dared to contradict the man who, more than most, controlled their purse strings and, by extension, their futures. Yes, carriers survived as the best politically engineered program in Pentagon history—more than two thousand companies in forty-six states performed more than $5 billion worth of work for each ship. Close to sixty thousand supply-chain jobs relied on the carrier construction. McCain held the navy leaders by their collective balls, and they knew it.

The admirals never needed McCain or anyone else to tell them they required a more modern carrier, and they planned to develop one. Shortly after the end of the Cold War, they envisioned new systems for launching and recovering aircraft and for moving weapons around the ship, making it a digital twenty-first-century vessel of war by focusing on the most critical components necessary for core carrier missions. They had started on the new carrier during the previous decade with a planned new combat system for the last Nimitz-class ship, the USS *George H. W. Bush* (CVN 77). It featured new sensor suites, better wiring, and other upgrades. They had planned to install a new major system on each of the first three next-generation carriers, what would become the Ford class, starting with a new aircraft-launching system on the lead ship, CVN 78. The admirals believed their *evolutionary* plan to be sound, measured, and logical.

Then things fell apart.

First, CVN 77's new combat system proved to be a non-starter. The company that was supposed to do the work, Lockheed Martin, privately blamed Newport News Shipbuilding, which, privately, blamed the navy, which, privately, blamed the contractors, as well as Congress and the general lack of resources. In the end, the carrier wound up with essentially the same combat system as the previous Nimitz-class ships. The disintegration of the rest of the navy's evolutionary plan lay squarely on Defense Secretary Donald Rumsfeld and the rest of the George W. Bush administration, which shifted to a *revolutionary* carrier modernization plan.

In that new post–Cold War reality, the Bush administration and the Rumsfeld Pentagon no longer needed to make incremental changes to stay ahead or keep up with the Soviet war machine. Now it could make transformational changes, take leaps, not steps, toward the future force. What better place to start than with the next generation of aircraft carriers, beginning with the future CVN 78? Rumsfeld agreed with the navy's decision to replace its heavy, messy, and labor-intensive hydraulic launch-and-recovery systems. He further agreed that the service should take advantage of the digital age to design ships that would be more electric. He saw no reason, though, why it should take three ships and more than two decades to do that. He wanted it all right away, on the very first next-generation ship.

Rumsfeld's approach turned out to be a recipe for disaster, but that would not become clear until much later. At the very beginning, Newport News Shipbuilding and the navy designers reckoned they were up to the task. They used the standard Nimitz-class carrier template. But inside, topside, and capabilities-wise, the new carrier would be a completely different ship.

They started with the reactors. Among other things, the nation's aircraft carriers served as floating steel homes for two nuclear reactors that, on their own, packed enough power to electrify the largest American cities, if not whole states. Each of the ten operating Nimitz-class carriers housed a dual nuclear heart: twin Westinghouse/GE A4W/A1G reactors, which powered four turbines and generated 280,000 horsepower. After about a half century of making, installing, integrating, and operating nuclear reactors on carriers and submarines, the navy and its shipbuilders had it down to a science. No US Navy nuclear-powered carrier had ever suffered a nuclear accident, and many in the Pentagon, and even Washington as a whole, knew no organization in the country or possibly even on Earth yielded more power than Naval Reactors, or NR, the division that oversaw all things nuclear for the service. The yard and the navy built the carriers, and various commands controlled the ships' operations through the decades, but Naval Reactors *owned* those ships. The officers, designers, and engineers there determined everything that would happen to that ship from conception through commission to decommission and finally to deconstruction. Some considered NR a cult.

To launch a new class of modern carriers with the latest technological advancements, a new reactor would be necessary. NR needed something smaller, lighter—weight being everything on a ship—and yet more powerful. It needed a more compact heart with a stronger beat.

NR found its answer with Bechtel Corporation's A1B reactors, made in its western Pennsylvania plant. Again, each ship would have two reactors, but although smaller and lighter than its carrier predecessor, the A1B generated a thermal power output of about 700 megawatts, or about 25 percent more than the A4W's 125 megawatts (168,000 horsepower) of electricity, plus 350,000 horsepower (260 megawatts) of thermal power. NR transported the reactors, about the size of railcars, in a way the navy would rather not be discussed.

A new reactor meant some design changes on ship. Specially developed nuclear materials aside, reactor assemblies were puzzled together the same way other assemblies were in the yard, at least at the basic level. Engineers redesigned and operationally "tested" the reactor assemblies in computer environments. The designers pinpointed, diagrammed, and calibrated pipes, wires, and valves, then pinpointed, diagrammed, and calibrated them anew to find the most efficient arrangement, which was then created as a single unit and, at the right moment, lifted and placed in the lower deck space of the carrier near the ship's middle. Special machinery in the shipyard's fab shop scooped out material from the middle of the plates to form a small bowl for the reactor to sit in. The carrier depended on those reactors to provide power for a quarter century. The navy classified the reactor systems' inner workings but relied on their proper installation at Newport News.

The navy and shipyard revamped its new Ford-class carriers to tap all that extra power. The redesign started with the sailors, whose berths would have better air-conditioning, more efficient lighting, and the first carrier installation of USB ports for cell phones, tablets, and other personal devices. Outside the berthing areas, the redesign made it easier to monitor machinery in the deepest, darkest, hardest-to-reach recesses. That machinery had to be checked and rechecked for efficiency and safety, by sailors performing those demanding and dangerous tasks. The new Ford-class design featured a network of sensors—cameras, microphones, and so on—and remotely controlled valves and other systems, cutting down on manpower and risk. To accommodate the additional electrical needs, Newport News also designed a new electric power distribution grid along the same basic operating principles as a city substation, but with enough power to generate 13,800 volts, or roughly three times the voltage generated aboard Nimitz-class carriers. Engineers created a whole new phone, internet, and electrical system for the carrier-city.

The navy had already put that additional power margin to good use. The most visible upgrade was the topside antenna changes on the iconic carrier island, with the elimination of most of those huge rotating radars Hollywood liked to highlight in movies featuring carriers. In their place on the *Ford*, the yard had installed a dual-band radar, or DBR, whose antenna was a solid panel the size of a building wall, embedded with small emitters and receivers to shoot pulses of radar beams on two different bands, boosting the ship's ability to conduct air-traffic control and to defend itself. That radar system emitted too much power for carrier flight operations, so the navy bought another, less-powerful radar suite, called the Enterprise Air Surveillance Radar, or EASR, to be placed on the *Kennedy* and later Ford-class ships.

Few things mattered aboard a carrier more than gaining better air-traffic control. But controlling aircraft wouldn't matter if the carrier couldn't launch and recover jets.

To those in navy aviation, carriers are "bird farms" for a reason. For the Ford class, the navy and its contractors had designed new Clancy-esque electric-centric systems. Even the elevators that moved jets from the hanger to the flight deck relied on electric components.

It's quite a thing to shoot a seventy-thousand-pound gassed-up fighter jet down a rolling runway into the air, time and time again, rocketing twenty-two tons about the length of a football field in about two seconds. US carriers accomplish this using a launching technique that dated back to the early days of the Wright brothers—harnessing the energy of a catapult that flings an aircraft with the force necessary to get it airborne. Of course, modern carriers rely on a more technologically advanced cat, as they are called.

During his *Ford* visits, Nasty Manazir recalled his own days in a fighter being catapulted off a carrier in one of those classic carrier movie scenes. He rolled his F-14 Tomcat—nowadays usually an F/A-18 E/F Super Hornet—to the green-shirted kneeling aircrewman crabwalking below the jet to connect the front wheels to a shuttle in the catapult groove running along the flight deck. The jet-blast deflector rose from the deck, a specially designed small wall of steel meant to absorb and deflect the sudden-launch exhaust. The shuttle connected to a metal slug below the deck, which in turn connected to a piston atop one of two parallel cylinders that ran lengthwise under the deck for each cat. Tapping the reactors, sailors filled the cylinders with steam to the pressure needed for launch, accounting for the type of aircraft, payload, wind speed, and other conditions. Done right, with the right amount of steam pressure deployed at the right time, and it was like a popped champagne cork blasting the shuttle and aircraft down the cat in a flash and off the flight deck in a cloud of steam.

Below the flight deck, the slug attached to the aircraft became a speeding tethered missile that needed to be stopped before it blew a hole in the carrier's bow. The navy used a technique like the barrels of water lined up on highways to stop a speeding car from crashing un-impeded into an exit divider, namely a water brake that needed about 1,300 gallons per minute to operate.

So much could go wrong on any given launch. Not enough cat pressure, and the jet would just plunge immediately into the sea. Too much pressure would mean a crushed jet nose gear, a scrubbed flight, possibly a totaled aircraft, and a potentially dead or injured pilot. Nimitz-class cats were limited in the size of aircraft they could launch, limiting the kind of aircraft that could be deployed with those carriers.

Navy mechanics found the legacy cats to be finicky systems, requiring teams of sailors to maintain the pistons, the cylinders, and the rest of the equipment in a grimy, greasy hellhole of a ship space.

Ship operators often took one cat or another offline for repairs during deployment. Some of the maintenance problems had to do with the navy aircraft—carrier planes were getting heavier. And the daily aircraft launches beat the hell out of the hydraulic cat systems. The equipment and spare parts also added several tons to a ship's weight. The navy needed a new launch system that weighed less, required less maintenance, and operated more reliably and precisely. The navy settled on the electromagnetic aircraft launch system (EMALS), developed by San Diego–based General Atomics, which replaced the steam pistons with linear induction motors, sending pulses or waves of electricity down the cat and pushing the aircraft down the flight deck. The energy for EMALS started in the same place as it did with legacy cats—in the nuclear reactors. Instead of building up cat steam, though, the energy was stored kinetically in rotors spinning at 6,400 rpm in shipboard generators bigger than the largest commercial ocean-shipping containers. When it was time to launch an aircraft, those generators emitted an energy pulse lasting two to three seconds to start the cat on its way, generating about ten million watts of electricity, or about enough energy to power a small town for those moments. The electromagnetic waves pushed the EMALS linear induction motors along a metal track about the length of a football field. The power and speed of those waves could be digitally controlled to launch different types of aircraft, unlike Nimitz ships, which could launch only the larger planes.

The only discernible launch difference as the Super Hornet or whatever roared off the bow of the ship was the lack of the smoke cloud after a successful legacy-cat launch. No pistons, no hydraulics, and no steam—few moving mechanical parts, significantly cutting down on spares, space, and weight. The new system also meant a major reduction in the number of sailors needed to maintain the launch system. As the EMALS wave could be halted as precisely as it had been started, the ship no longer needed the water brake, all those hydraulics, the immense water tank, and accompanying equipment: another reduction in weight, wear, and manpower.

Linear induction motors and the associated technology for EMALS had been around for a while, but putting it aboard an aircraft carrier would be novel and challenging. The impact of a slight construction variance here or there in the installation of one of the legacy Nimitz cats likely would not have a major impact. However, with electric-wave-producing magnets, precise alignment meant everything, and it would be up to the yard to redesign the ship for the new system and then install it.

In the spring of 2012, shipyard crews had boarded the ship in the cool of the dark night and climbed into the trough, a shadowy cavern that stretched into the blackness at the end of the ship. It was the same size as on Nimitz-class ships. A spring breeze brought aboard some of the cool damp air hovering above the James River. Under a temporary, slanted aluminum covering that protected the in-deck EMALS installation, steelworkers broke down the three-hundred-foot-long launch line into twenty-nine segments and then balanced laser trackers down the trough to align the magnets along the laser's path to within seven-and-a-half thousandths of an inch. They accounted for corrosion, a serious concern on any kind of surface on a ship that would spend its life in saltwater, which could knock the alignment off by as much as three-eighths of an inch along the whole line. The backbreaking, tedious work took all night to complete. When they finished, they reviewed their work—a trench wide and deep enough to hide a small car that ran more than half the length of the flight deck. Rows of small panels, about the size of large flat-screen TVs, with rows of electrodes sticking out like little fingers, ran up to the bow of the ship on slightly slanted trench sides.

Launching different and more varied squadrons of aircraft faster and cheaper would mean little unless a carrier improved the way it landed the planes. The launch required brute force and power, but the landing, or recovery, counted on a maritime aerial ballet, a choreography of ship, wind, and aircraft that ended in an always dangerous controlled crash. No matter how many times navy and marine pilots "trapped"—landed a jet on an aircraft carrier—every time put all the cockpit senses on red alert. Most times, they circled the rolling, pitching ship to size up the situation. Then they set their approach from the stern, revving up, slowing down, dipping wings this way and that way, moving their feet and hands to maneuver the aircraft controls for that perfect line of attack as the postage stamp on the ocean from miles back grows into the shape of an aircraft carrier, with not-yet-discernible runway lines on the deck.

The aircraft approached from the carrier's aft, looking like a movie in slow motion, as the ship headed into the wind to provide a little extra lift for the wing-rocking jets as their noses inch up in the final descent, their tailhooks hanging like enormous bug legs. Then, suddenly, everything sped up. The pilots throttled to full military power—just in case this controlled crash wound up being a bit less controlled than everyone liked—and *smack!*—the seventy-thousand-pound plane skidded down the flight deck with a force that shook the decks below, its tailhook grabbing one of the metal wires as thick as a sailor's arm stretched across the deck from two white metal cylinders like giant retractable guitar strings. If the tailhook caught no wires . . . well, that was why the aircraft was revved up to full power for that moment, so the pilot could lift those wings and bolt to the sky once again. The carrier aviation club called them bolters. Navy pilots used the word as a verb, too, as in, "I boltered." Carriers designed the landing runway at a fourteen-degree angle to the rest of the ship, so bolters avoided hitting anything as they took off again.

In a successful landing, the tailhook grabbed the wire, pulled it tauter than anyone but an engineer could imagine, like an overwound guitar string, and the aircraft slammed to a screeching, lurching stop—looking until that last halting moment as though it would break that cable and hurtle off the deck. The cable slackened and dropped back to the deck as the green shirts rushed out with what looked like upside-down brooms and started "sweeping" the cable back into a straight line as it was retracted back into the metal cylinder. In the berthing spaces below, it sounded like a thousand ghosts clanking their chains all at once.

While the skill, practice, and verve of the pilots and deck crews all worked together to stick the landing above deck, the machinery and mechanics made that possible all day along, all night long, every day, every night. Like the legacy launching operations, the dirty, dangerous, and dull work required much manpower, with systems taking up space and weight the navy wanted to eliminate. It took quite a few sailors and a powerful bit of hydraulic machinery to play out 374 feet of wire, one and seven-sixteenths inches in diameter.

Below the flight deck, the heart of those operations—an arresting engine the size of a locomotive—circulated 380 pounds of hydraulic fluid at a pressure of ten thousand pounds per square inch to bring a thirty-five-ton fighter traveling 150 miles per hour to a stop in 315 feet in two seconds. These arrestors took up too much space, weight, and manpower. The navy wanted a more modern—more electrical— setup. Again, General Atomics provided an answer: the advanced arresting gear, or AAG, which replaced decades-old technology of hydraulics-based machinery and manpower with new tech built around an electric motor. Instead of that massive hydraulic ram, the AAG used rotary engines spinning water turbines—"twisters"—at ultrahigh rates to absorb the force of those jets stretching those cables. Again, the navy relied on shipbuilders to figure out a new design-and-integration plan for the ship.

Now, standing on the back end of the *Ford* on the Tidewater waterfront in the spring of 2014, Nasty Manazir envisioned the squadrons he could deploy with his new launch-and-recovery systems—F-18s and the new super-stealthy next-generation F-35 combat jets, as well as large drone aircraft the ship could send out to much greater distances than it could currently, to spy or bomb. All this would make the vessels less vulnerable to attack from enemies in the South China Sea or the halls of Congress.

AAG and EMALS provided another benefit, as well: boosting the launch and trap rates would speed up operations in general, what the navy called the sortie rate. The faster the sortie rate, the more effective a carrier's aircraft in combat. Navy and shipyard designers searched for other ways to boost that rate, and they found them—ranging from rather simple in-deck refueling stations to more exotic advanced weapons elevators (AWEs), which moved more weapons faster from the belowdecks magazines to aircraft being readied for launch.

Before the in-deck refueling station design, Nimitz-class sailors refueled aircraft by stretching large, heavy fuel hoses thick as a weight-lifter's legs across the deck. During that time, the carrier could neither launch nor recover any aircraft, cutting significantly into the sortie rate. Designers thought, why not move the refueling operations closer to the aircraft to speed things up and make the job a lot safer? The *Kennedy* and other Ford-class ships featured six in-deck refueling stations at the front and back of the flight deck, where crews could pop a square deck hatch about the size of a small car door to hook up the planes for more gas. Pilots just taxied over, topped off, and went on their way.

But pilots needed more than fuel to continue combat operations. They needed to rearm with bombs, missiles, and other payloads. Deck crews needed to reload the planes before relaunching them; otherwise, it would be like sending a racing car out of a NASCAR pit stop without a new set of tires. To better protect the ships, the navy located the magazines for such weaponry deep down in the bowels of the ship, bringing the bombs to the deck in massive weapons elevators. Nimitz carriers had nine of them, each of which could carry 10,500 pounds—

more than the weight of two SUVs—at a hundred feet per minute. These were not your usual office-building floor climbers. They were the size of truck loading-dock platforms, with large metal railings on three sides. Unlike office lifts, operating along their own shafts running unobstructed through buildings on the edges of floors, these elevators ran *through* the floors—the decks. Rather than having a single shaft that would enable a fire to spread from the flight deck down to a magazine, designers offset the elevators into upper- and lower-stage segments. The upper-stage elevators do not go any lower than the second deck, while the lower-stage elevators do not go any higher than the hangar bay.

When the air ops really hummed, sailors used the crew's mess near some of the elevators to stage bombs and other weapons before moving them on a lift. Then, on the deck, sailors stored the weapons on the "bomb farm" before loading them on the aircraft. The system had done the job for decades, using the same pulleys, wires, and hydraulics to move those huge platforms through bulkheads, decks, and compartments via a system of hatchways and other access points designed to open before and close watertight after each ascent and descent. It would be like doors suddenly opening in the floor of some office building, a platform full of bombs zipping through the room, then the floor closing and then doors in the ceiling opening for the elevator to continue its ascent.

But the navy carrier planners wanted something faster that offered greater power and efficiency. They wanted to replace all those pulleys with a modern electric system that would require less space and maintenance. Carrier designers also planned to improve elevator operations by bringing the weapons all the way up from the magazines to a dedicated staging area. Modernizing the elevators would also increase the carrier's sortie rate.

The shipyard had been building and installing Nimitz-class elevators for decades, so there was no reason for the company engineers or the carrier overseers to doubt Newport News Shipbuilding's ability to create a satisfactory new weapons lift. Yard designers came up with a plan for eleven elevators—two more than the *Nimitz* had. Each lift accommodated twenty-four thousand pounds—more than twice the payload capacity on Nimitz, enough for four SUVs and a VW. Those elevators zipped along at a speed of 150 feet per minute, 50 percent faster than their predecessors. The new elevator setup also eliminated the need to use a bomb farm or the crew's mess. Unlike launch-and-recovery systems, the new elevators would be the sole province of the shipyard.

Yard designers replaced the legacy pulleys and machinery with electric actuators. The navy required that land-based prototypes for the new electric-centric launch-and-recovery system be built and installed at the service's aviation testing center in Lakehurst, New Jersey, but there was no similar requirement for the new elevators. The navy discounted any need for such a backup.

EMALS definitely did need it, however. Making all the components work together as a system to simply launch an aircraft proved to be not so simple. It took some coaxing to make the system work at all, and even then, it often broke down. To troubleshoot, crews needed to "spin down" the electric EMALS motors and generators, which took an hour and a half.

Fixing EMALS was expensive. In his Pentagon office, Nasty Manazir fielded a request for $50 million one month and another $50 million request just a few months later. He felt like screaming "Fuck You!" into the phone. The carrier community had never needed to deal with this problem before. Weapons elevator installation and operation had been seemingly set on automatic for half a century. Sure, electric motors might be new on a carrier, but industry had been using them for years. Why the hell couldn't these contractors, these navy engineers, these waterfront workers get this new launching equipment to do the one damn thing they were designed specifically to do?—launch an airplane. Hell—launch anything.

They needed another $50 million. And another $50 million. And another. As the *Ford* price tag climbed over $12 billion and fell under greater scrutiny of the likes of McCain, EMALS became the bane of Manazir's existence. He hated going to the Hill like some kind of used-car salesman. The *Ford* was not a used car.

The EMALS failure to launch caused enough headaches, but then Manazir also dealt with problems with the advanced arresting gear aircraft-recovery system. "A software issue!" Nasty remarked to the carrier program office. "What software issue? Oh my God—you've got to be kidding me." Sometimes, the conversations turned a bit nastier—and saltier. The AAG could not handle an off-center aircraft hit on the wire. The software was not designed for the AAG to compensate for a tailhook grab to left or right of center. The AAG didn't "understand" how to "re-center" the airplane. Then the problem arose with the water twisters, which required more than a few trips for Nasty from the Pentagon to Newport News.

Manazir and the yard experts studied the twister turbines encased under the *Ford* ship deck and shook their heads. As it turned out, bringing a screaming jet to a sudden stop caused more strain—more torque—on the gears than expected, and the fasteners used to anchor the water twisters could not handle the additional stress.

"That twister is our biggest risk," he reminded the shipbuilders during one of his visits in the spring of 2014. Proven technology and simulations sometimes just failed to cut it. The navy, shipyard, and General Atomics pooled their mental resources and came up with a redesigned fastener. This required more than a software fix. They had to replace hardware on the existing ship components and systems, and then test the *Ford*'s operational equipment, which was risky and expensive—more time, more money, and more angst. Especially back in DC, as Nasty would find himself once again in the hot seat on the Hill. It didn't help that Pentagon testers found reliability problems with the AAG that, like those with EMALS, were hard to fix because of the electrical design. While the navy carried the responsibility to furnish the launch-and-recovery systems, until the yard delivered the *Ford*, the carrier remained the shipyard's problem.

What started as a transformational experiment in the early days of the Bush administration became something of a financial and programmatic albatross in its latter days, carried along with the mounting war costs in Afghanistan and Iraq. Military spending shifted further away from the naval needs of a Cold War to the ground forces and intelligence resources required to battle terrorists. As the *Ford* price tag broke the $13 billion mark, carrier opponents gained ground, particularly when it seemed the ship failed to even reliably launch and recover aircraft. Throughout the latter Obama years, CVN 78 became a harder sell, and the whole class, if not the entire US Navy carrier fleet plan, seemed to be in jeopardy. Navy Secretary Ray Mabus and Admiral Jonathan Greenert, Obama's chief of naval operations, or CNO, bragged more about their new little warship—the littoral combat ship (LCS)—a vessel somewhere between the size of a PT boat and a destroyer, with the firepower of neither. Compared to the carrier, it generated as much respect as a bathtub toy. While combatant commanders—the military officers charged with protecting US and allied interests in various theaters around the world—clamored for more carrier deployments, the CNO claimed their new little warships would be the best weapon to thwart Chinese aggression. Meanwhile, China progressed with its own aircraft carrier program. What would China be like with a carrier navy, carrier proponents asked, if it was already giving US the naval equivalent of the finger?

Forget LCSes—admirals and other US combatant commanders throughout the world wanted more carriers. Many others in the navy, who believed carriers were the alpha and omega of US naval operations, felt the same. Carriers cut a large swath across different navy communities, most of which held a lot of power, not only in the navy and Pentagon, but also in the halls of Congress.

Shipbuilders in Newport News, of course, saw no sense favoring LCSes over carriers. The Obama administration's fawning attitude toward the LCS began to feed the worst fears about the president and his Pentagon among some of the workers at the yard, while the union itself supported the president and his administration's prolabor policies. Obama was proving to be a divisive element along the waterfront. They had expected big things from the president. After all, his wife, Michelle, agreed in 2012 to sponsor the new attack submarine *Illinois*, which was then under construction.

Several months before the *Ford*'s launch, on a gusty, warmish wintry day at the end of February 2013, Obama visited the yard and spoke about the impending Congressionally mandated sequester that would freeze budget spending and imperil carrier and other large defense programs. With a giant ship propeller standing behind him in the Supplemental Modular Outfitting Hall and looking all business in a dark suit, Obama addressed the throng of applauding shipbuilders.

The applause started moments into his speech, when he said, "I usually don't get a chance to hang out with nuclear submarines, especially submarines that my wife has sponsored." And it grew when he quickly added, "It's a great chance to see the incredible men and women who, every single day, are helping to keep America safe and are just the bedrock of this country's manufacturing base. . . . The main reason I'm here is to call attention to the important work that you're doing on behalf of the nation's defense, and to let the American people know that this work, along with hundreds of thousands of jobs, are currently in jeopardy because of politics in Washington."

He was preaching their gospel.

"In a few days, Congress might allow a series of immediate, painful, arbitrary budget cuts to take place—known in Washington as the sequester," he warned, putting the entire onus squarely on lawmakers. "Instead of cutting out the government spending we don't need— wasteful programs that don't work, special interest tax loopholes and tax breaks—what the sequester does is, it uses a meat cleaver approach to gut critical investments in things like education and national security and lifesaving medical research."

Amen. Even if the steelworkers didn't actually say it, the feeling spread through the crowd.

"The sequester will weaken America's economic recovery. It will weaken our military readiness." *By taking away American carriers and subs, taking away our contracts.*

Obama seemed to be reading their minds: "The threat of these cuts has already forced the navy to cancel the deployment or delay the repair of certain aircraft carriers. One that's currently being built might not get finished. Another carrier might not get started at all. And that hurts your bottom line. That hurts this community."

Hurts? More like kills.

Then Obama segued into what he *really* came to talk about—his $930 billion spending-cut plan, which included a $580 billion plus-up, thanks to the elimination of tax loopholes and deductions he termed "wasteful."

Republicans had made it clear they wanted nothing to do with the proposal. Obama addressed them: "My door is open. I am more than willing to negotiate. I want to compromise. I don't think that's too much to ask. I do not think that is partisan."

He stopped for a moment's applause. "The majority of the American people agree with me. The majority of Newport News agrees with me. We need to get this done."

Another pause for more applause.

"And the other thing we've got to do is to stop having these crises manufactured every month," he said, then added with a bit more conviction, "I know you guys must get tired of it."

A longer pause for much greater applause.

"If it's not a good way to run a business, it's sure not a good way to run a country."

Amen.

The president had said all the right things. Among those praising the president's speech was United Steelworkers Local 8888 president Arnold Outlaw, who noted that worry over what Congress would do had actually created a bit of détente between the union and management in recent contract negotiations.

Obama had made Congress the devil, and many union steelworkers were ready to cast lawmakers to the nether regions for the precarious position facing the yard.

Some shipbuilders, though, believed the president needed to shoulder some of the blame for the state of affairs.

One thing nuclear pipefitter Dan Sunderland noticed when he started working at the yard—a bit before the *Ford* launch, when work started to ramp up on the *Kennedy*—was the growing number of Obama shirts, bumper stickers, and similar tokens of support outside or even inside the shipyard gates.

Sunderland had moved into Newport News from Ohio to attend X-ray tech classes at Riverside Hospital, just up the road from the yard, and he'd stayed on at Riverside before joining the shipbuilder ranks. He climbed down into carrier and submarine nuclear engineering spaces to test the pipes' ability to maintain the high pressure required by those nuclear systems. He preferred going into the subs, which, when connected pierside, would be nice and cool in the air-conditioning. But no amount of AC seemed to make it comfortable in the carrier. You always sweated, especially in those engine and auxiliary rooms. He turned up the pressure on those systems and watched it rise, feeling the sweat bead on his forehead and roll down his face. One thing he liked about the carrier, though—there was some room to walk around. In a sub, you had to crawl around or bend over like a human question mark. That was the hardest part of the job—crabwalking around, dripping sweat and just dealing with the daily ship environment. Well, that and dealing with management. Sometimes the rules they enforced . . . they just seemed silly. As a nuke, he often had to wait for the mountain of paperwork and approvals to be OK'd and processed. He'd find himself a bucket to sit down on and a bulkhead to lean against. But management didn't like that. No sitting, they told him. But what could he do while waiting for the OK to start his job? Was it his fault they hadn't planned this better, to get the task-sequencing right? What could he do when he didn't have all the parts or kits he needed? "Well, if you have to sit," they said, "at least do it with a drawing in front of you so you look busy."

He had been a steelworker back in Ohio. There he learned that when you came to work and put on your steel toes, you were going to work. No one waited around. He saw those master shipbuilders at the yard now, with four decades of experience, and he told a reporter, "They really don't know what the outside work world is like, what normal jobs are like." He wondered if it would be different if the yard had to truly compete for business and had to turn a profit. Look at the way the yard got navy contacts. When you buy a Chevy, you wait for autoworkers to finish it, and then you pay for it. When the navy buys a carrier, though, it (that is, the taxpayer) pays for it basically by the worker by the hour—hour after hour. So the yard got paid for the hours workers were there, even if they had nothing to do at the moment. At least that's the way Sunderland saw it.

But all of that was above his pay grade. As was the political climate in the yard when he started there, just as Obama was running for his second term. He'd tell anyone who asked, "I don't care what somebody is; I care who they are."

Obama was, in his opinion, an apologist who had sold out the US to other countries. Under his administration, the nation had dropped in operational carrier strength, and the yard had laid off workers. Sunderland didn't see anything Obama had done for shipbuilding or for the country. Yet everybody in the union here seemed to love the guy. If he had to be honest, every one of the Black members in the union.

When Obama came through the area, making appearances, officials at the steelworkers Local 8888 were pushing tickets—"Here's your ticket! Here's your ticket!"—to make sure everyone had a chance to see him. The union wasn't like the one they had at the Goodyear plant in Ohio. Sunderland just couldn't understand why this local was so Democratic, so pro-Obama. As he'd tell folks, "I'm White. I'm Christian. I'm pro-life. I'm pro-gun. I'm not apologizing. I am who I am. I'm not asking anyone else to apologize for anything they are. But I've been a Republican my whole life. It just makes more sense to me. I don't want to insult your views. That's just how I feel."

Sunderland was voicing the way many older, White workers at the yard said they felt about the leadership at Local 8888, whose presidents have been mainly Black for decades. Black workers saw nothing wrong with that. Many of the Newport News Shipbuilding steelworkers were Black, after all. For about a century, Newport News Shipbuilding had been operated and managed exactly as it was—a huge industrial plant in the South, a fine setup for White workers. Work hard, learn your job, and do it well, and not only could you be guaranteed a waterfront wage for the rest of your working life, but you could advance. You could become a foreman, or even more. No matter what, you could earn the kind of living that could raise a family. That's what it was all about for most of the (mostly) men who entered the yard gates day after day during those years. But Black workers saw it differently—as another world. Those who worked there before steelworkers gained their foothold saw it as a plantation.

The yard mirrored the rest of Southern factory-like business at the time, according to author W. J. Cash, who wrote, "The Southern factory almost invariably was . . . a plantation, essentially indistinguishable in organization from the familiar plantation of the cotton field."

Newport News Shipbuilding was unlike other yards in one key

respect—the Tidewater waterfront had always been a work draw for the Black population. Other yards had seen migrations of Black people and other minorities during World War II. No sudden rush of such workers flooded into Hampton Roads to cause workplace upheaval. That meant that what local historians had called a paternalistic system at the yard continued during those years largely unchanged. Trades officials trained the apprentices they wanted to, which meant other White men. World War II did open yard gates to a significantly larger number of women workers, but it would take until the 1970s for the first female to graduate from the Apprentice School.

The shipyard continued to draft Black workers for the yard's filthiest, most menial, and inherently most dangerous work. Most of them simply relished the opportunity to have a job. They earned more than they could make tilling the land, working the water, or toiling in some factory. Those who didn't like it could just clock out and leave for good. Plenty of other Black men waited to take their place.

This one-yard, two-worlds setup continued into the 1970s, until those two worlds began to see the benefits of a single combined existence to battle a single foe—shipyard management. Those good wages simply didn't match those of other naval shipbuilders around the country, let alone compare with the salaries of other Tidewater-based industrial companies. You could earn more making beer for Budweiser up in Williamsburg. Forget about Black skin or White skin—what mattered more was the blue collar. As the Obama presidency was coming to a close, the *Ford*'s problems mounted, the schisms in the yard deepened, and workers began to wonder if the union could still be the uniting force it had been before. So much was at stake. The nation's carrier program. The yard's future. Their jobs. They all hinged on getting the *Kennedy* done right. Could the waterfront workers at Newport News Shipbuilding put their differences aside as they had done before, when they joined forces to toss out the company-stooge representatives of the Peninsula Shipbuilders Association and embrace the steelworkers, surviving the company goons, the state police batons, and their own prejudices to do so?

The seas sure seemed stormy on the Tidewater waterfront. But, as Wilbur Wright once observed, no bird soars in a calm.

3

STEEL VETS

Known to some by the nickname Castro because of his penchant for quoting Malcolm X during the turbulent 1960s, Bill Bowser seemed the most unlikely of possible union organizers. Compact, nimble, and wiry, he spoke in a surprisingly commanding voice, which he used to hammer home point after point after point. Bowser seemed a more likely candidate, by his own admission, to wind up dead in a bloody race riot. Oddly, he became a union and shipyard legend exactly during that kind of action. No other single worker did more to bring the steelworkers union into the shipyard.

Like most who earned their bread on the Newport News waterfront, Bill Bowser grew up the son of a shipbuilder. His father, Wilburn Bowser, worked as a coal trimmer and timekeeper for the C&O railroad, before moving to Newport News from North Carolina in the middle of the previous century, where he started in the shipyard as the supervisor of a cleanup crew, then joined on as a shipwright, shouldering enormous hull plates into place.

One of the numerous Carolina men to make the trek to Newport News in search of a shipyard job, Wilburn Bowser represented exactly the kind of worker Collis P. Huntington had in mind to float his new venture when he wandered the James River waterfront after Reconstruction in his salt-and-pepper beard and mustache, planning operations for the C&O and the future shipyard.

Huntington's yard continued to build that workforce even as it lost money on its first ships and had to rely on the Huntington family railroad fortune to keep it going. The Spanish-American War fanned the US desire to grow a strong navy, a need outlined by Alfred Thayer Mahan, president of the US Naval War College, in his 1890 book, *The Influence of Sea Power*, which has served as a bible for American, and later Japanese and Chinese, naval development and shipbuilding right up to today. The book underscored a simple premise—navies not only protected nations; they also protected national global shipping lanes. Maintaining those commercial lanes, establishing strong trade, and guaranteeing the ability to sail where and when you wanted to—that was what fueled a nation's rise and helped it maintain power. The need for a strong navy and the US global naval force later anchored by its aircraft carriers hinged on that single selling point.

Huntington focused on building good navy ships even as the yard lost money doing so. More than ships or even a company, he desired to build a community. The company continued to donate land for playgrounds, schools, a hospital, the first public library, and a spot that would be named Huntington Park. No one who lived or worked there at the time could tell where the company ended and the community began, even after Huntington died, in 1900.

The US entry into World War I brought about the next set of big changes in the shipyard, particularly at the very top. Newport News Shipbuilding president A. L. Hopkins was killed aboard the Cunard liner *Lusitania* when a German U-boat torpedoed and sank the ship in May 1915. His replacement was Homer L. Ferguson, who would ensure the shipyard's survival through two world wars and a depression, saving the yard in the 1920s when both naval and commercial contracts dried up.

Born in the North Carolina Great Smokies, Ferguson found his calling in shipbuilding, first as a navy constructor—later known as the supervisor of shipbuilding—and then as the shipyard president. Unlike previous, absentee chiefs who managed the yard from afar, Ferguson lived in Newport News, setting the precedent for all succeeding yard presidents. He used his proximity, constantly patrolling the waterfront inside the gates. Lean and short—"terrier-like"—he wandered about, chatting with the workers, his "aggressive" brown eyes staring out from horn-rimmed glasses sitting on a big nose. He often talked out of the side of his mouth. Having forbidden smoking inside the yard, he ducked outside the gates to toke cigars, the smoke curling around his head and shock of gray hair. Ferguson continued to channel the Huntington spirit, funding civic projects and charities. Everyone knew that if you wanted anything done, in or out of the yard, you had to tell Mr. Ferguson.

Following the 1922 International Disarmament Conference, the navy canceled $70 million in contracts. Newport News became as quiet as a cemetery, snagging whatever work it could find throughout the decade—building barges, dredges, tugs, yachts, and passenger vessels, building and repairing locomotives and railroad rolling stock, and constructing hydraulic turbines and other heavy machinery. The yard produced turbines, gates, pumps, pressure regulators, trash rakes, valves, and other equipment for hydroelectric power projects, including the Hoover Dam and the Grand Coulee Dam, the largest hydroelectric power project in the US, not to mention a wind tunnel for Hampton Roads' Langley Field.

While the rest of the country roared during the 1920s, Newport News Shipbuilding barely creaked by, regaining its stride after the stock market crash with the buildup toward World War II. The yard saw its future as the US naval shipbuilder secured with the 1942 Battle of Midway, which anointed the aircraft carrier the premier naval warship. With its massive dry dock, ready workforce, and decades of experience, Newport News Shipbuilding appeared tailor-made to be the carrier cradle through the Cold War and later decades. In 1948, Congress funded the biggest carrier ever planned, a sixty-five-thousand-ton behemoth supercarrier that would be 1,090 feet long, capable of launching four-engine bombers carrying nuclear weapons. It was to be called the *United States*. The yard laid the keel for the ship in April 1949 with little fanfare and enormous hush-hush. Within days, though, the Pentagon canceled the order, but later carriers carried nuclear weapons through the mid-1990s.

The Soviets tested their atomic bomb in August 1949, and in June 1950, Communist forces invaded South Korea. Suddenly, Washington lawmakers willingly opened government checkbooks for plenty of carriers, and the yard prepared to build three of the enormous warships straightaway.

In August 1957, the navy announced its intentions to build a real nuclear-powered carrier. Newport News laid the keel for what became the *Enterprise* in February 1958. While the yard had been building carriers for decades, the nuclear ship required a whole new set of procedures, materials, and training protocols. The yard introduced new construction methods in welding to build with sophisticated new alloys and novel techniques in constructing nuclear-powered machinery. Welder qualifications took on a whole new meaning. Overseeing all this, Rear Admiral Rickover demanded strict adherence to detail, spelled out in written procedures along these lines: *Pick up wrench with right hand. Hold nut with your left. Turn wrench.*

In the fall of 1957, the Soviets launched and orbited Sputnik, turning up the heat to get *Enterprise* out to the fleet. The navy commissioned the ship on November 25, 1961, two days after Thanksgiving, and Newport News declared it Enterprise Day in the city.

With the carrier work building up, the yard, city, and region anticipated a hefty bump in the workforce and accompanying boost in the local economy—the beginning of halcyon days. But social upheavals in the country began to upset the old order in the South, in Tidewater, and in the yard, causing some angst among Black workers and the overall yard workforce. Newport News Shipbuilding had more Black workers on its payroll then than any other company in the country, but those numbers never amounted to any true labor bargaining strength. Not that yard workers had any real clout as a unit. Indeed, the Peninsula Shipbuilders Association, or PSA, which "represented" yard workers, often operated as more of a help than a hindrance for management. Yard workers tried to form a proper union, and the US Supreme Court in 1939 had upheld a National Labor Relations Board order to stop the yard from interfering with those efforts. In the South, though, labor unions received as warm a welcome as Union forces did during the War of Northern Aggression.

In Virginia and North Carolina, dirt-poor farmers like Wilburn Bowser, Black and White, happily accepted any decent-paying jobs. The elder Bowser entered the yard as a rigger, a job he held for more than two decades. The backbreaking, spirit-sapping work required him to wrestle with steel slabs the size of small cars. At that time, the elder Bowser and other men of color likened themselves to serfs toiling in a gated kingdom as they trudged through yard gates to the waterfront or one of the brick buildings by the river. Black workers drank from "colored" fountains and used piers instead of bathrooms to relieve themselves. In the South, they had conditioned themselves to accept such second-class treatment. At least here in the yard, they made decent money.

At that time, management and workers gave safety short shrift. Foremen chased safety inspectors back to the gate. There were no safety lines to grab. The men didn't care. They went into the yard to make a paycheck to feed their wives and kids. They needed nothing more. They made enough for a home, a car, and some pocket money. For guys who hadn't finished high school, that was the good life. Some of the guys couldn't even write their names on their time cards. They didn't have Social Security cards, just a bunch of little copper tabs with numbers on them.

But they worked hard for that money on the waterfront. Men moved metal, mountains of it. Propellers as tall as a two-story house—just get skids, rubber, and grease, and muscle them on. No forklift around? Just make yourself a "Georgia buggy"—rig up something with two wheels and push. Riggers could hang a fifty-ton wheel on a cradle of chains just right, move it around, and settle it in just so. A normal work week was 112 hours, 16 hours a day every day, including weekends.

The significant Black population in Newport News and the rest of Hampton Roads serving in the yard created a community almost like a southern Harlem, with an electric nightlife to match. The Jefferson Theater and the other hotspots near Two-Five and J—Twenty-Fifth and Jefferson—were decked out like Christmas all year round, the signs flashing above the well-worn sidewalk. White folks called it lurid. Cab Calloway, Bo Diddley, Sarah Vaughan—all the greats came to play. The city had its own circuit—Chestnut Avenue, Newsome Park, Phoebus. But uptown, on Jefferson, that's where you found the action. Wilburn Bowser's son, Bill, loved that action, and missed it terribly when he did his stint in the service, stationed in Savannah, Georgia, in the early 1960s. Once home, he returned to the Newport News jazz scene—and followed in his father's footsteps in the yard.

Bill started there in 1964, working on the bull gang, scores of Black men grappling with miles of cable thick as a firehose on the carrier deck, pulling and pushing the lines meant to carry power and communications through the ship. He worked that detail, wearing thick gloves, the fingers covered in heavy industrial tape to cover the holes torn by handling the cable—chanting call-and-response songs that harked back to slavery days, his arms so weary by the end of the shift, he could barely lift them to grab the steering wheel of his car. But he could make fifty bucks a week and enjoy his weekend nights at the Two-Five and J.

Wilburn Bowser wanted his son to start and spend a career at any place other than the shipyard. The elder Bowser wanted more for his boy, and he rode Bill mercilessly, day after day—hour after hour sometimes—to make more of his life. He wanted Bill to take advantage of all the money he had coming through the GI Bill and go to college. The younger Bowser did not share his father's dream and wanted nothing to do with more schooling. Bill tried to have it both ways. He asked for an education leave from the yard, working during school breaks with a job waiting for him when he finished. He knew plenty of White workers who requested and received the same thing every year. The yard turned him down and laid him off. The dutiful son took his GI Bill money and drove off to the North Carolina Central University at Durham, a historically Black school, to study accounting and become a CPA. But Bill Bowser hated Durham. It offered no Two-Five and J. It had no Jefferson Theater. It had no soul. He spent more time studying *The Autobiography of Malcolm X* than *The Elements of Accounting*.

Malcolm X wrote, "I would rather have a brother with a processed head than a processed mind." Bowser didn't let those professors process his mind. He saw college education as a sellout to the White man's definition of success. He wanted to go back to Newport News and become the shipyard's Malcolm X. Back home during a school break, cruising the familiar Hampton streets next to his father in Wilbur Bowser's Pontiac Bonneville, Bill sniffed in the new-car smell, bracing himself for the conversation destined to end badly.

"I just don't think college is for me. I want to work a trade. I'm going for a job back inside the yard."

"Boy, I don't want you working there."

"It's where I can get a job."

"Then get out of my house," said the father. In the summer of 1968, Bill Bowser left his parents' house and went back to the yard. He started working in submarines, squeezing his body into the tightest of crevices to muscle up sound-dampening panels for the nation's stealthiest warships of the deep. Small Black men like him seemed destined to be stuck with the worst jobs on the piers and in the ships. "Tarred and fettered"—that's how Bowser described a day in the deepest bowels of the new Los Angeles–class attack subs, cutting and pasting acoustic-dampening panels on the inside of the sub hulls.

Nothing taxed his frame more than pushing the panels around, Bowser thought, as he wrestled another piece into place in a space as dark as the sea and as narrow as a turnstile. But the glue—a sticky paste that looked like tar and smelled like rotten eggs—it coated everything. Fingers, hands, arms, neck, head, hair, face, shirtsleeves, shoes, socks, trouser legs. It never came out. He felt it, smelled it, and tasted it long after he made it home, showered, and slumped in his easy chair.

Back he went every day to do his job. He and the rest of the army of men like him, tunneling into the darkest sub spots. Down into the sub. Squeeze into place. Measure, sometimes standing on studs. Cut out a plywood template, twelve-foot by twelve-foot. Tack in place to make sure it fit. Take it down. Lay on the vile glue. Cut out the panel. Wrench into place. He understood the importance of the work—not only for the yard, but also for the navy and the country. Workers, though, reviled the job as one of the most awful in the yard—and one, Bowser observed, imposed on few White workers. He turned to his colleagues as they submerged for another day's work and stated the obvious. "It's just brothers. That's all we're good for." They had no control. Anger rose within. He filed and won a grievance about the unfair work distribution, but nothing really changed. Of ninety-six men working in the subs, the White workers numbered only six. Of eighty-three supervisors on the waterfront, only one Black. Bowser tried to climb the ladder to a better job. He stopped one day to check out the sub plans in the shop. A White foreman asked him, "What are you doing?"

"I want to learn."

"You don't need to know anything about this."

Other waterfronts operated the same way. Craftsmen in trades at other yards around the country did the same, training promising White men in the trades. When Bowser complained to higher-ups, they told him, "You had better do as you're told. That man has your job in his hands." Do as you're told. Climb back into the subs.

Those subs cost about half a billion dollars each, and each one earned the shipyard about 18 percent profit. While the yard would come to be more associated with the iconic aircraft carrier, the submarine contracts helped keep the waterfront employed during the carrier off years. Building the subs struck workers as something akin to sailing on them—the claustrophobia on the boats made being in them very difficult and, to be honest, drove builders and submariners nuts.

Big Ed Elliott had worked on subs throughout his career. Like Bowser, he had worked the relatively small 688 Los Angeles–class SSN attack boats and the behemoth SSBN ballistic missile submarines, the ones that carried nuclear weapons. But Big Ed worked as a machinist—still a body-twisting job on a sub, but not as onerous as Bowser's glue job.

Spending months at a time in any job underwater with more than a hundred others crawling through narrow halls of steel, electronics, and stale air has always exacted a physical and mental toll. Building these ships—the responsibility to make sure these hollow tubes of metal nuclear-reactor casings operated for decades beneath the waves without fail—also weighed heavily on those charged with the task, whether they glued panels like Bowser or machined torpedo tubes, like Big Ed. They never thought of sending sailors undersea in subs if the workers harbored any uncertainty about the subs' safety. Big Ed and other sub builders climbed into a sub before it launched to bounce it around, checking for leaks, and woe to any rookie working with them, who needed to be christened. The foreman would sneak a few squirts from a water bottle to scare the newbie. One time, they had a kid in there, a VP's son, just out of college, and they showered him, yelling, "We got water! We got water!" The kid came up from the bilge, panicking, panting, and shaking.

Besides playing with rookies sometimes belowdecks, the sub builders spent some of their time topside shooing away seagulls that had an affinity for the periscope, antennas, and the area around them on the mast. One day, one of Big Ed's team climbed, leopard-like, up a periscope while a seagull perched atop. He grabbed the bird by the legs, climbed down the ladder, and proceeded to head off the boat toward the end of the pier with the full-grown, squawking, ticked-off bird.

The foreman stopped the gull-hunter, asking, "What in the world are you doing?"

"This is my pet now. I'm taking him home."

"You let that seagull go right now. I mean you let it go right *now*!"

And he did. They watched the bird fly off, giving them all the what-for.

Newport News shipbuilders had a thing about seagulls.

There was one foreman whom others found, well, irritating. He always kept a bag of chips or some other snack in his desk, carefully "hidden" under some paperwork, that he always dipped into every day around midmorning, making a trip back to his office for a quick bite. Some other foreman caught an injured seagull who was unable to fly—and they stuck it in that drawer a few minutes before the expected snack time. True to form, the foreman came in, opened the drawer, stuck his hand down—and the bird went, as those present attested, *nuts*.

Such stories unfolded whenever veteran shipbuilders like Big Ed, Wingnut, and the Murphys would gather and reminisce. None of them doubted, in their later years, that nowadays they would probably face reprimands for such "inappropriate" behavior.

It was that kind of behavior, though, and thousands of other little idiosyncratic yard behaviors, that united the workers, made them into a family of sorts. When women workers came into the yard in greater numbers, the menfolk watched where they, the men, dressed and undressed. No longer could they just drop their drawers wherever it was convenient. The men saw the women coworkers as . . . well, like sisters—or, in other cases, something more. The yard became a place to start a family, as workers courted behind the gates, at yard sporting events or functions. Family members helped one another, and yard workers operated on that principle, inside or outside the yard. If someone was sick, everyone donated. They all worked hard, did the best they could, and held their heads high as each ship was delivered.

That sense of family began to wear away at Bill Bowser's militant views. He began to view all the waterfront workers as brothers and sisters. The former disciple of Malcolm X started to see things through a much different prism after working for years in the yard. He realized all workers shared a common enemy—company management. The real fight in the yard was not between Black and White: it was really between hard hats and suits.

Of course, shipyard management saw it quite differently. Yard leaders enjoyed a casual, loose relationship with the PSA. There was a Southern gentleman's sense of respect for one another. Management's overall view of labor relations was best summed up in a *Fortune* magazine piece from the 1920s, which said the yard's "attitude toward its employees is that of a paternalistic seigneur" with no room for an "outside labor union."

In his seminal 1941 book on Southern attitudes, *The Mind of the South*, W. J. Cash noted the incorrectness of using the word "paternalism" to describe the relationship, saying, "It suggests . . . there existed on one hand an essential dependance, and on the other a prescriptive right," or, as he put it, "The right of the . . . master class, to ordain and command." As far as labor organizations went, Cash wrote, "Unionism was Com-mune-ist, and so a menace to their Southern heritage."

There was little hint of overt racial strife at that time in the yard or at Newport News. Black workers simply did not have the opportunity for better jobs in the caste system of Tidewater, which hometown author William Styron described as "segregated and a world apart."

That started to change in the 1960s in Newport News, as it did throughout the country as both labor and racial tensions bubbled to a boil. Desegregation did not sit well in Tidewater, and groups like Save Our Neighborhood Schools (SONS) protested, holding signs saying BUS THE JUDGE. The unrest hit home in the shipyard when the Reverend J. Cornelius, a Black crane rigger and also president of the Newport News chapter of the NAACP, and three other shipyard workers filed a lawsuit charging the company with violating the Civil Rights Act of 1964, saying Black workers made less money for equal work and could not secure promotions. The NLRB and EEOC investigated. At the time, Black workers accounted for only 32 of 1,997 supervisors and only 6 of 506 apprentices, the workers who were often marked for future management jobs. A couple of years later, the yard reached an agreement with the government to level the playing field for its Black employees. Even though management called the allegations "hogwash," the company did not want to put any of its navy contracts in jeopardy. That agreement, though, did later lead to over 3,800 Black promotions by the 1980s, as well as openings for supervisory positions for 99 Black workers, and apprenticeships for 421 Black recruits. Another 785 received extra education to move up in the company.

While Black employees started to make strides in the yard hierarchy, the overall workforce also decided to show a little muscle, too. The PSA voted to strike in 1967, the first serious one in the yard's history. The move shocked management, which claimed the strike breached the bonds that had traditionally existed in the yard. Workers stormed the gates in a midnight riot on July 9, and Virginia State Police restored order, essentially ending the walkout. Workers returned to work after one short week, gaining nothing.

As the PSA licked its wounds, the global conglomerate Tenneco bought Newport News Shipbuilding. Tenneco planned to build a small fleet of tankers to carry Alaskan oil that it appeared the country needed to survive the international energy crisis. Workers hoped that a multinational company would break the parochial plantation ways they battled along the James River, but in the first contract Tenneco negotiated, the workers lost sick leave.

The sudden onset of a worldwide energy crisis, an Arab oil embargo, and soaring oil prices torpedoed Tenneco's plans for that tanker fleet. Tenneco suddenly had no justification for a larger fleet of oil tankers, and the increasing cost of energy meant each ship would cost more to build. To make things worse, the country's shipyards—including Newport News—went to court with the navy to see who would pay those increasing costs for military contracts. Tenneco had to bail out Newport News to cover expenses.

Through all this, PSA continued to "represent" the workers. Hoping to goad the association from within to take a harder stand against the yard, Bowser became a PSA delegate. But no one there wanted more trouble. Bowser knew he and the yard needed professional labor help—a tough task. The company had done everything it could to trip up union efforts in the yard. When the designers wanted to organize, they lost their jobs; making the union stamp was as good as a pink slip in the eyes of most. Still, working as a machinist in 1969, Bowser started talking to the International Association of Machinists about organizing the trade workers at the yard. The IAM was a part of the AFL-CIO, a real union, Bowser thought. The IAM could change some things—if Black and White workers joined forces. After seven years, though, the IAM had gained no traction in the yard. In October 1976, his IAM brothers told him of another plan—to bring in the United Steelworkers of America to organize the whole yard.

The suggestion dumfounded Bowser. "Steelworkers?" he asked. "What do they know about shipbuilding? They work in the hills."

You're mistaking them for ironworkers, his IAM brothers told him. Give them a chance. He and a handful of Black workers met with steelworkers during the late afternoon in a ground-floor room in the Hampton Holiday Inn. Very casual—jeans and clean shirts, top buttons undone. "I'm a machinist," Bowser reminded others there. "What am I doing here?" But he started to change his mind. The steelworkers had more than two million members. They had high national standing. They fell under the AFL-CIO umbrella.

They also wanted to organize everyone at the yard. Their success promised to open doors for him and other Black workers to move around to different jobs—or better yet, move up to better jobs—much more easily. Previous piecemeal organizing efforts all had failed, with different organizations or unions vying for different kinds of workers. But if the steelworkers organized most of the rank and file under one banner, they could negotiate deals for the workers to climb their way up. That caught the attention of Bill Bowser, a college-educated man on track to learn other trades and skills, like working with lasers. The IAM organizers officially relinquished their effort, and the battle began to bring in the steelworkers.

A war like this forges some odd alliances. Like that of Bill Bowser and Russ Axsom.

Axsom, a welder who had joined the yard just a few years before Bowser, came from Virginia mine country, where labor ruled like a religion. Axsom joked that the picture he stared at as a kid on his living room wall was not that of Jesus, but of union icon John Sullivan. On Bowser's walls hung portraits of Malcolm X and Martin Luther King Jr., as well as a tricolor flag—black, because that's the color he was born; green to represent the military service to his country; and red to represent the blood and pain of the movement. Bowser and Axsom had one goal in common: "All that matters, Russ, is that we get the union in here."

"Bill, I'm with you."

4

LABOR GAINS

Their different skin colors meant nothing to Bill Bowser and Russ Axsom. They shared blue-collar goals, and that mattered most. They stood side by side in front of the yard gates, talking with the river of workers as they filed into the yard in the gray-blue dawn before the first shift, about the absolute need for union representation. "Together, we can make a difference." The couple traveled to places like Gloucester County, talking up a union. There, locals gave Bill Bowser reason to be apprehensive, if not scared. Folks on the other side of the York River greeted uppity Black and White union organizers with equal contempt. One day, a deputy in those parts gave them the age-old Southern warning: "It'd be best if you weren't here when the sun went down."

Bowser's father issued his own warning about the likely company response to all this union stuff: "They'll get you—they'll get me."

But once he started, Bowser couldn't stop. He talked union all the time now, down at Vernard's Barbershop in the Newport News East End. In the bars. At the Waffle House. The organizers enlisted the help of the local churches, the centers of Peninsula life, as civil rights leaders had done decades earlier, as when Doctor Martin Luther King Jr. spoke at the city's First Baptist Church in January 1958. Newport

News Black folks had a long tradition of being religiously raucous, going back to the 1930s, when Bishop Charles "Daddy" Grace led rousing revivals near the yard and Elder Lightfoot Solomon went from selling seafood from a cart in the city's North End to being a radio evangelist—one of the first—with his *Happy I Am* program, which CBS claimed had twenty-five million listeners.

At Bowser's church, the First Baptist Church on Twenty-Ninth Street in the East End, the Reverend W. M. Brown Sr. thundered on about the union in true Southern Baptist fashion. The workers, he said, needed to organize. He implored congregation members to attend nightly meetings in the neighborhood about the organizing. Of course, other churches in other communities delivered a different sermon: Stop the union!—a message much more in line with traditional thinking about organized labor in Tidewater. Unions went against the grain, they said, the dues eating up part of your paycheck with little return on the investment. The idealistic nature of unions didn't help put food on the table or money in your pocket.

The yard laid off Bowser and dismissed his father in 1977 as it hit a small trough in navy business. The yard usually operated that way, laying off when the work dried up a little, hiring back when flush with contracts. The younger Bowser kept up the organizing, risking a chance to get his job back. One thought consumed him—unionizing the yard.

Finally, in January 1978, the hard work paid off when 9,093 of the yard's 17,000 hourly workers voted to replace the Peninsula Shipbuilders Association with the steelworkers' union in an election overseen by the National Labor Relations Board. That night, Bowser and the other organizers screamed themselves hoarse at the victory party at the Plumbers and Steamfitters Union Hall on Warwick Boulevard. They hugged, kissed—even cried—as they congratulated each other with soda, chips, and sandwiches. Bowser and a few others walked next door to the exotic-dancing club for some harder celebration, and the party didn't end until after two in the morning. Bowser needed surgery for his throat—months of talking and cajoling, not to mention the night of celebration—had exacerbated a polyps problem. His voice retained a raspy edge from then on.

Later, having regained his job, Bowser told his fellow organizers. "Now the hard part starts." Having wrangled enough votes to have a union, they now needed to sign up the actual members. "And you know the company will be watching our every move," Bowser said. Organizing on company time could be a firing offense. *But we have a chance*, Bill Bowser thought. *Finally, a chance for dignity—with United Steelworkers of America Local 8888.*

Newport News Shipbuilding lost a court battle to have the vote disqualified but still refused to recognize the union. Bowser, Axsom, and the other activists built up the membership. Soon they had signed thousands of workers. Local 8888 called for a strike in January 1979, a year after the workers voted in the steelworkers.

The yard figured it needed about six thousand workers to keep going—and it expected many to cross the picket lines. That didn't happen, as workers put aside any racial, geographic, or other cultural schisms to form one blue-collar front under the steelworkers banner. Local cops lined Washington Avenue in front of the yard gates. State troopers rode into town to join them. Workers marched and picketed as the police—some tugging on leashes held taut by jumpy dogs— eyed them with suspicion. Tension blanketed the streets like heavy August humidity. Workers lining up the pickets at the Thirty-Seventh- and Fiftieth-Street gates talked of the 1967 riot, wondering if the same violence would occur now. Bottles sailed through the air and shattered across the street. The workers chanted "Steelworker power!" and "We're number one!" and "We shall not be moved." Bowser implored his union brothers and sisters to keep their cool and not provoke the authorities by doing something stupid. On April 13, thousands of workers poured into the modernistic Hampton Coliseum, with its concrete crown of lighted outside walls resembling church stained-glass windows when lit at night, to hear the company's latest "offer"—a demand that the workers unconditionally return to work immediately. The rank and file refused.

Intermittent ragged clouds spanned the sky on April 16, 1979, an uncharacteristically cool spring afternoon in Tidewater. Outside the yard gates, though, tempers rose and soon exploded in what they referred to from that moment on as Bloody Monday. With the state troopers there and the governor apparently backing the company, the tension deepened like a coming summer storm. Bowser managed union business out of an office in the Williams Building near Thirty-Seventh Street on Warwick. One of his brother yard workers limped in, bloodied. "The cops are beating up everyone on the block," he said. Bowser rushed outside to the first police officer he saw, asking that the beatings stop. "We'll get people out of the street," he said. He walked down Washington to Thirty-Third Street to the union's strike headquarters across from the iron shipyard gates. There, police and union members now clashed in a melee of fists, feet, and nightsticks. Bowser raced down the street; it was a war zone. A white van pulled up, and a handful of police officers jumped out. One swung a three-foot baton into Bowser's left leg, and he crumpled to the ground. Police descended on him, kicking and clubbing. He folded himself into a fetal position and lay there, refusing to fight back. He retreated into the fortress of his own thoughts. *Be like Martin Luther King Jr. Nonviolence. Nonviolence.* The beating continued. His body went numb. *Please stop. Martin Luther King Jr. Nonviolence. Please make it stop. Make it . . .*

As EMTs carried Bowser to an ambulance on a stretcher minutes later, he raised a fist in the air. He took days to heal and suffered a permanent limp. About a week after Bloody Monday, the shipyard dropped its unconditional demand for workers to return to their jobs. The union told its members to return to the yard and let the company's appeal work its way through the courts.

A federal appeals court upheld the union's right to recognize the workers, and in March 1980, the steelworkers signed their first contract with Newport News Shipbuilding, the workers ratifying the agreement on March 28 by a vote of 4,939 to 1,646. The contract provided salary increases ranging from eighty cents to a buck per hour. Workers had their first real contract, and Bowser told the new members that that mattered most. For the first time, welders, riggers, and other blue-collar workers had true representation—a union. Steelworkers, every one of them. "Things are going to be so different now," Bowser told Axsom. They felt that way not only about the yard, but also about the whole South. If the United Steelworkers of America could gain a foothold in the biggest private employer in Virginia, anything seemed possible. In much the same way that freedom riders and civil rights marches had ignited passions against racism, the union victory in Newport News could spark a new labor uprising all over the South. "We have to do whatever we can for the movement," Bowser told his new brother and sister steelworkers.

Bowser won a union officer's seat. He studied by-laws, contracts, and any kind of labor material he could find. After the day shift, other steelworkers found him in the squat new brick union hall on Huntington Avenue, holding court on rights and benefits. They all knew him, the guy with the red suspenders and union buttons, carrying a green canvas knapsack. A combination medicine cabinet and tool kit, his pack contained dental floss, Band-Aids, New Skin, Tylenol, other pain pills, a rain suit, a flashlight, coveralls, soapstone, grease pencil, screwdriver, and an extra pair of glasses. No one expected anything less from a guy who had fifteen sets of work clothes in his closet, numbered in the back to keep them rotated in proper order. He possessed seven pairs of work boots.

From the steps of the steelworkers' hall, across a desert of parking lots and asphalt, Bowser eyed the shipyard office building as the stronghold of the enemy, an executive's fortress of bricks and concrete. Now the workers yielded power, too. "We have a union now," Bowser reminded everyone. "If we stick together and follow the proper procedures, we can hold this company accountable."

Now the new union local needed to push the labor movement to the limit and do battle with the shipyard. With the arrival of the eighties, Newport News Shipbuilding received its biggest boost in defense contracts in decades. Ronald Reagan charged into Washington with a mandate and a plan to make America strong again. Reagan wanted a six-hundred-ship navy and soon bought two Nimitz-class aircraft carriers, bringing with them thousands of additional workers—and more potential union members. Bowser and Axsom made the rounds, bringing as many as possible into the steelworkers fold. The union planned to do more than command a movement; it wanted to build an army.

However, traditional mind-sets died hard in the South. A groundswell of *worker* opposition, holdovers from the old days, tried to decertify the yard in 1983, but 75 percent of about eighteen thousand eligible workers voted to retain the steelworkers. Nevertheless, by 1988, the steelworkers had signed only a third of the eligible workers. About a year later, fourteen union members, most of whom sat on the local grievance committee, wrote to the United Steelworkers of America (USWA) international headquarters in Pittsburgh, warning about a possible local collapse. Some said the local leadership focused on too much politicking and too little hard union business. Some workers even talked of bringing in the Teamsters.

Then the Berlin Wall came down in a scene of jubilant celebration, signaling the collapse of the Soviet Union, the end of the US navy shipbuilding boom and guaranteed fat contracts for the Newport News shipyard, and casting doubt on the Tidewater steelworker workforce and union power. No longer would Newport News welders be putting finishing touches on attack subs. The Soviet maritime threat had all but vanished, and Washington torpedoed the Seawolf sub program, killing the most lethal, capable, sophisticated—and expensive— subsurface fleet ever put to sea. Once again, the Beltway set its sights on the carrier program. Fewer people saw a need for the multi-billion-dollar warships bristling with weapons for wars no longer on anyone's radar. Newport News Shipbuilding needed to find a way to survive without relying on more overly generous big-bucks contracts from the navy. In 1994, the yard turned toward its greatest and most pliant and resilient resource, its workers, for a lifeline.

"They want us to do *what*?" Bill Bowser asked new Local 8888 president Alton Glass at the Union Hall. The company, Glass said, wanted the steelworkers to give up vacation days, forgo pay raises, and drop requests for better pensions. Otherwise, the company might not be able to give the navy the kind of deal necessary to secure funding for the aircraft carrier CVN 76, later to be named the *Ronald Reagan*. Without that carrier, Bowser and other union leaders knew, they needn't worry about getting into any harsh negotiations with the company. There wouldn't be a company to negotiate with, at least not the vaunted carrier builder on the James.

Glass rode to Washington to lobby for the carrier. Bowser helped persuade yard steelworkers to give something back to keep their jobs—a tough sell in a place where the jury still deliberated on the value of the union and a graph of union membership numbers resembled a roller coaster. During contract time, the union roster usually swelled. Union leaders wanted as much power behind their punch as they could muster, so they stepped up their organizing efforts. Workers, for their part, seemed to forget about the union between contract negotiations. Most let their dues lapse, until the next contract membership drive started its usual ascent. What also helped cut into the union roll was the laying off and rehiring of workers to match the yard's ebb and flow of work. Oftentimes, workers searched for more reliable work. Those who joined the union one year, then, might not even be at the yard the next.

The "Castro" Bowser of decades ago would have rejected the company concession requests outright. But he now realized that when it came to union business, you didn't try to get what you wanted by any means necessary. Sometimes you took a step back now to take two forward later—even if that meant a salary freeze, one lasting five years.

The company secured the *Reagan* $2.5 billion construction contract in December 1994 as the union membership provided some negotiating room for the yard, agreeing to give up vacation days and agree to pay and pension considerations. The union expected the company to cede some of the concessions back to the steelworkers after the contract award. Instead, new company president and chief executive officer William Fricks sought private shipbuilding contracts on a global scale while cutting costs on navy deals. The Fricks management team said Tenneco prevented the yard from meeting the steelworkers' requests.

By the mid-1990s, Tenneco regretted its Newport News acquisition. The Texas conglomerate spent $68 million for robotic steel fabrication, upgraded computer networks, and other enhancements to help the yard win the *Reagan* contract and compete for international commercial shipbuilding deals. Despite the upgrades, the future appeared grim on the waterfront for both commercial and additional navy work. Clinton defense secretary Les Aspin considered alternatives to carriers for an overseas presence, possibly going from the carrier fleet of thirteen at the time to as low as ten. Fricks knew such a cut threatened the company's ability to keep its gates open. While building new carriers no longer guaranteed a viable business, refueling existing nuclear ships did, so the yard fought like hell to secure the USS *Enterprise* overhaul, a contract that helped keep the yard as well as the ship afloat.

Tenneco spun off the yard, and Newport News Shipbuilding went public in December 1996. As Fricks told his employees, "Officially, this is our shipyard."

Bowser read the notice of the NNS public offering in the company newsletter, *Yardlines*, and again felt some hope. "Now, they'll be able to negotiate freely," Bowser told other union members. He needed to believe things would get better, that all his work mattered. Fricks set other priorities for his solo company than give-backs for union workers, like figuring out what mergers or acquisitions might be in the best interests of him and the company.

With another union contract coming up in 1999, Bowser obsessed over organizing. They needed to raise those membership numbers, and he worked night and day to do so, phoning folks, knocking on doors, talking up the local. His doctor warned him about his smoking, high blood pressure, and constant stress. He took separate pills for anxiety, insomnia, and headaches. He found it just as stressful to maintain a union local as to create one. Virginia was a right-to-work state, so all yard workers received the same union-negotiated deal regardless of who had officially signed up as steelworkers. As the 1999 negotiations started—thanks largely to the efforts of Bowser and other organizers—the union rolls swelled to almost 100 percent. "People here just don't realize what they have here in this union," Bowser told brother organizer Wayne Deberry. "They just don't know what they have—or could have."

"I know it, Soldier."

Deberry had given Bowser the nickname Soldier out of a respect that bordered on idolatry. The younger steelworker knew of the war wounds Bowser carried from Bloody Monday. Bowser represented the real deal, the union front line. Deberry had earned a place in Bowser's heart as well. Like Bowser in his younger days, Deberry became an elected officer, a grievance committeeman, who also knew the union contract by heart.

Local 8888 leaders were now determined not only to regain what they felt they had conceded in the '94 deal, but also to boost pension and pay scales. They found that particularly difficult now, three years later, as the company faced takeover bids, failed commercial attempts, and a shrinking federal bankbook. Newport News Shipbuilding and United Steelworkers of America Local 8888 faced each other—and stalemated. Labor and company leaders met for preliminary luncheons, but the hard hats believed the suits were trying to intimidate them. The rank and file particularly wanted pension changes. Many steelworkers who had worked in the yard for as long or even longer than Fricks faced a pension of less than $1,000 a month. In their sixties, some limped to work like wounded war vets, but could not afford to retire with the current average monthly pension of about $500. A worker at a yard up north received more than twice as much. It galled the union even more because the company had a $150 million pension-fund surplus, earning $15 million each year in interest.

Yard workers in the shop received about fourteen dollars an hour; workers at the Ford plant in Norfolk made twice that. Workers wanted real bonuses. Executives felt the workers wanted too much in tough economic times. Steelworkers of America Local 8888 went out on strike for a second time in the spring of 1999.

Steelworkers staunchly backed Local 8888 president Arnold Outlaw, a mountain of a man whose soft, cool voice belied his huge Black frame. Some might mistake his soft-spoken manner for indecisiveness or a lack of conviction. They quickly discovered their error. He entertained no intention of backing down on his local's demands. Like Bowser, he hoped the company could be more flexible after it got out from under Tenneco's thumb, but yard executives claimed they could not cover union-demand costs as a solo company.

Shipyard executives said publicly they expected the strike to last no longer than two weeks. It dragged out for four months, with Bowser leading the charge on the front lines every day. Bowser felt no fear pacing picket lines in the same streets where police had beaten him two decades earlier. Tension still stretched nerves, but everyone seemed calm—picketers, police officers, company officials. Walking the streets in his white T-shirt with UNITED WE BARGAIN, DIVIDED WE BEG printed on the back, Outlaw noted more police than picketers at most spots, telling the media, "They've got nothing to worry about from us." And Bowser did not necessarily see that as a good thing. Workers marched, chanted, and blew whistles, but he sensed resignation in the air instead of outrage. On the street, Bowser actually saw steelworker brothers and sisters chatting and smiling with police officers on the picket lines, a sight that angered him. "We are at war!" the veteran steelworker warrior told his protégé, Deberry. "And the picket line is our M16."

Work stalled on the carrier *Reagan*. Strike resources dried up, including those bankrolled by the USWA. Outlaw seethed as USWA leaders told him they could do nothing more for him and his strikers. Local 8888 took its campaign to Richmond, to make its demands known at the company's annual stockholder meeting. There, church leaders led striking steelworkers along the streets, holding up signs with slogans like TORPEDO INJUSTICE! CLASS WORK DESERVES A CLASS CONTRACT! and DON'T LET BILL FRICK US. Jesus, they said, wanted workers to get a fair pay. Bullhorn in hand, Leo Gerard, the international secretary-treasurer, told the strikers that the company contract proposal sent a strong message: "Southern industrial workers live in poverty, and we want to keep them there." The yard, he said, wanted to keep workers racially divided. It was time, he said, to unite.

Waterfront workers never lost sight of the fact that minorities filled less than a fifth of official and managerial positions at the yard, while Black members comprised only a quarter of the corporate board of directors. At the same time, the constant playing of the race card caused ill feelings among some White steelworkers, who felt a bit left out, as if the Black workers thought they rated union recognition more than anyone else. The feeling festered over the following decade.

The strike proved expensive for steelworkers—the international paid out about $700,000 per week just to give 8888 members money in their pockets.

Finally, union and management, frayed and frazzled, reached a tentative agreement. Bowser took the agreement home and stayed up all night, listening to the Duke and underlining in pink and yellow marker details of the pact. He knew in an instant where to look in the previous contract for specific grievance language that could have bearing on the deal. He didn't like what he saw.

The workers secured their raises—more than three bucks an hour on average, as well as their vacation days and better pensions. They negotiated better guarantees for seniority recognition and disability protection. But they gave up so much. Look at the length of the deal: Sixty months! He shook his head and forced a swallow. Five years. No one agreed to such a long contract these days. They needed flexibility so they could coordinate their contracts with navy carrier deals—the company wanted to avoid a strike during the middle of carrier construction or refueling. The deal also made it easier for the company to hire outside workers or use other ways to force workers off the job. And still no dental. Bowser battled with the contract all night long. By daybreak, the day of the vote, he decided to vote yes, mainly because of the increase in monthly pensions—by the end of the contract, they would be up to $900.

At the union hall, others disagreed. Arguments broke out. Shouts rang, inside and out. Deberry lobbied loud and hard to vote down the deal. "The language is too vague," he said. "It leaves us open to too much." Bowser agreed with the criticisms, but they needed to compromise this time. "He reminds me of me from before," Bowser told others. "He's always out there. He's not afraid to say anything to anyone. He cares about this union. Sometimes he goes too far. Sometimes he needs to restrain himself."

The strike fund money dwindled. They had gotten the pension increases—a major goal this time around. The company also boasted of the 18 percent increase in wages over the term of the contract, saying its workers would, on the average, make at least a dollar more than the state average pay for a manufacturing worker.

Two thirds of the membership agreed with Bowser, one of the first back through the gates that first week of August. But stress, tension, and work had done their damage. A week after the strike ended, he had a heart attack and needed a quadruple bypass. After about a year, he went back to Steel City to work in the North Yard in the briny breeze of the Tidewater spring. He didn't care that the place had become a pond after a heavy rain, with some holes as deep as a man's knee. He was back, his heart attack restrictions over; he returned to more strenuous duties in the North Yard that spring in 2000. He and his partner were cutting out metal vents roughly the size of large television sets to be placed in the carrier *Reagan*.

Reagan became a bone of contention between the union and the yard. During the strike, Newport News put its managers to work in the ship to keep getting navy checks, and the union didn't appreciate the fact that the government, in the local's collective mind, underwrote the yard's strike costs. Payoffs for layoffs, steelworkers called it. When the strike ended and the steelworkers returned to the job, they said they had to rip out and redo the supervisors' poor work. It took fifteen to twenty years' experience to perform some of the work those managers attempted, some of them with only fifteen to twenty hours of experience with the tools.

Workers also said the yard rushed them through some of the work to catch up, requiring more welding than normally would be required for a carrier, adding weight to the ship, and improper welds likely led to operational and maintenance problems later. Besides, those welds did not look like Newport News welds.

Bowser now focused on pushing, pulling, and cajoling *Reagan* vents into place. The awkward pieces weighed thirty to forty pounds and usually required two people to maneuver them safely. Bowser's partner went off for a bathroom break. With managers pushing workers to do everything as quickly as possible to make up time lost to the strike, Bowser decided to attempt to install the vent himself. He wrestled with it to shift it a couple of inches to make it fit properly. It slipped from his hand and fell from his waist to the ground, its pointy edge slicing through his left foot.

Bowser pushed the rehab and hobbled back into the yard after a few weeks. They put him six stories up in the *Reagan* island house. Up he'd climb, in the middle of summer, limping in his heavy work boots, limping through the passageways, crying in pain. The union, the brotherhood, proved to be almost worthless now. Legally, contractually, the company was within its rights. Union officials said there was little if anything they could do.

Bowser hoped for better luck with the new yard owners after Northrop Grumman bought the yard in the spring of 2001. But, again, Local 8888 steelworkers found, when it came to unions, the new boss ruled in much the same way as the old boss. When the Newport News shipyard ran short on electricians—a claim Local 8888 disputed—Northrop flew in workers from its other yards. Deberry filed grievances on the outside workforce—and then talked to the press about his concerns. Deberry felt the union needed a push—and he figured the media could be his leverage, but Bowser told the younger leader, "You shouldn't put union laundry out in public."

Still, Bowser acknowledged, another union-company battle began to brew under the Northrop banner. For the first time in his life, Bowser—now fifty-nine years old—thought he might not be up to the fight. The foot refused to heal. He went back to work with an official medical recommendation for light duty for thirty days, no standing or walking for more than two hours at a time, and no carrying anything over twenty-five pounds. Nothing helped. The doctors tried more therapy. More drugs. The pain proved unbearable. Bowser could no longer work. Pain medication now made it difficult for him to concentrate on even a simple conversation. Doctors who had been treating him since the beginning said he had no chance of reclaiming his old job. The yard demanded second opinions from other company-picked doctors—who said there seemed to be no reason why he couldn't go back to work. The company offered Bowser a settlement of $65,000 to finish his career at the yard and take him off the rolls. No medical. He sat in his living room chair and listened to the Duke, watching the sun filter in on some days, the rain smacking the panes on others. He had waited for decades for the company to bestow upon him a sense of dignity. Now he knew real dignity came from within. He would not be returning to work on the *Reagan* or any other carrier.

The yard wanted to finish the *Reagan* as quickly as possible, but that proved more difficult than anticipated. The navy had marked the ship as the first of the new carriers with incremental technological changes toward the next-generation flattop that would in a few years become the carrier *Ford*. Specifically, CVN 76 would be the first carrier built with a new shipyard-developed internal communications network—the Integrated Communications and Advanced Networks, or ICAN—which would tie together all on-ship communications and certain control systems using modern digital equipment and technology.

ICAN development followed a trend in navy ships noted by shipbuilders for years—a tendency to make the vessels a great deal more complex. On earlier ships, the fire main system consisted of a pump, a pipe, a valve, and a sailor. Now the fire main system featured a variable speed pump and a remote-control valve, with sensors all over the ship to tell you whether to open or close the valve or to turn the pump on or not. The navy and shipbuilders had invested money to replace sailors with sensors, and that alone made the ships more sophisticated—and more difficult and expensive to build. ICAN was a prime example. The navy wanted a way to make all its disparate systems and networks communicate with one another—even though it would be like connecting modern computers via two cans and a string. As the prime contractor, Newport News Shipbuilding would be spearheading an effort to make that work, starting with ICAN.

The yard proposed ICAN as a building block of sorts, a foundation for the next carrier change that would take place—a revamped combat system on the next ship, the *Bush*. The next carrier—known at the time only as CVNX-1—was to feature the electromagnetic aircraft-launching system. The next ship, CVNX-2, would include the new high-tech aircraft-recovery system, and then the next would sport a new radar as part of the planned next-generation carrier evolution.

Reagan became the first carrier built with ICAN added in the construction contract, while the *Nimitz* became the first carrier to get the system put on board during its midlife refueling, another project at the yard. Shipbuilders included ICAN in the initial designs for carrier CVN 77, which the yard started to build in 2001.

Navy carrier officials started to rename the Newport News system the I CAN'T. The *Nimitz* combat systems officer reported fundamental problems with the technology as soon as the ship went to sea, saying it failed "support mission requirements" and represented a "significant step backward from legacy standards." The ship couldn't even get its public address system to work. The combat systems officer reported, "All observations indicate a dramatic reduction in ICAN system reliability as circuit usage and numbers increase." Headsets designed specifically for ICAN "fail at an extraordinarily high rate." No off-the-shelf replacements were available and "in fully manned watch situations, critical stations are working without headsets," the *Nimitz* message said. The problems had been previously reported but no viable long-term strategy for correcting them had been developed, the message said, adding that *"Nimitz* is concerned that the full impact of ICAN shortcomings are not fully understood by NAVSEA Program Office."

Worse, long-term impacts of the ICAN problems, including those connected to the proposed new combat warfare system on the carrier *Bush*, were directly related to later issues with the *Ford* and all the tension in the designing and building of the *Kennedy*. The navy awarded the yard a contract in January 2001 to build the *Bush* for $3.8 billion, which included delivering the ship's warfare system, a job previously handled by the navy. Newport News hired Lockheed Martin to help with the development, procurement, and integration. The navy and yard had figured on spending about $450 million for the new combat center work.

Bush warfare center problems started in the summer of 2000 when unplanned additional expenses associated with the combat center started to force the navy to shift money and other resources there. ICAN and the *Bush* warfare system were woven together. The same technology that made ICAN work allowed the warfare system to coordinate nine capabilities, such as missile tracking and launching, and to integrate these functions with those of other ships, creating, for the first time, a combat system that gave the world's biggest warship some of the same tracking and missile-firing capabilities as a destroyer. However, some of the problems with ICAN generated issues with development of the warfare system, and it was already pretty much impossible to decouple the ICAN and warfare-center problems at the electronic DNA level. When US lawmakers offered to provide another $250 million to fix the warfare center, the Pentagon rejected the new combat system as not worth the investment.

The navy decided to equip the *Bush* with the same battle system that other carriers had employed for about four decades, and shipyard engineers and navy planners needed to find a way to reconfigure it with the ship's ICAN components. Congress warned that the change created possibly greater technology risks—and higher costs—for CVNX, and these fears proved to be prophetic.

When the George W. Bush administration started in 2001, the Pentagon chief nixed the carrier evolution and instead called for a revolution, killing the CVNX plan and planting the seed for what would become the CVN 78 Ford-class carrier.

Before the yard and navy considered focusing on the *Ford*, they needed to make up all the strike-lost time on the *Reagan* and deliver a Newport News–quality ship. Lee Murphy, Big Ed, and the other seasoned shipbuilders saw the goal as a matter of pride. Just before the *Reagan*'s christening in the spring of 2001, as they prepared to flood the dock, Murphy walked down to the ship. He stood alone and eyed the huge bulbous bow protruding from the rest of the hull like a miniature submarine. The design kept the big aircraft carrier from nose-diving as she crashed into waves head-on at high speeds, shoved along by her four twenty-one-foot, thirty-three-ton propellers. But a mini-sub affixed to the front end of one of their carriers—man, that was pretty cool. Soon this hull would touch the sea, and the bow would be beneath the water. And a Mathews man was a general foreman of the men and women who built this ship. "This ship is mine," he whispered. "This is my ship, because of the job we've all done." And he beat it out of there as fast as he could, before someone saw the tear running down his face. Carriers would be the closest that most steelworkers would ever get to reaching some kind of immortality.

On a wet, blustery March 4, 2001, the navy and the yard christened CVN 76. Sitting in the audience, Local 8888 president Arnold Outlaw said, "That ship is union-made."

The future of the world and the carrier-building program changed again following the attacks on the World Trade Center towers and the Pentagon on September 11, just a half year after the christening and launch of the carrier *Reagan*. The Bush administration shifted all focus, resources, and concern to Afghanistan and then Iraq, with little bandwidth for future carriers.

Costly warfare-center snafus and construction delays plagued CVN 77, exacerbating doubt not only about the *Bush* but also the next-generation Ford-class carriers. Yard management called in some of its experts, who not only helped finish CVN 77, the last Nimitz-class boat, but also went on to spearhead work on the Ford-class ships.

Newport News shipbuilder Sam Carper had just finished managing back-to-back engine-room overhauls for carriers *Washington* and *Vinson* when the call came from his director on Memorial Day weekend in 2004 as President George W. Bush was gearing up to campaign for a second term. "I have some good news, and I have some bad news," Carper's manager told him. "The good news is, you are a wanted man; the bad news is, I am being asked to send you to CVN 77. I'm going to have to send you over there to work the night shift and help get the engine rooms and propulsion plant ready to start their test program."

Carper returned to the bowels of the ship, climbing those dirty bilges and handling those filthy valves. But this time, he did his job on a newly built ship instead of revamping systems and equipment on an overhaul ship. Everyone knew what a problem the *Bush* had become, even beyond the issues with the warfare center. "Well," Carper told his boss, "if that's what I gotta do, that's what I gotta do."

That had been his mind-set since coming into the yard at age eighteen, about a month after graduating high school in Covington, Virginia, a small paper-mill town near the state line with West Virginia. Another Apprentice School grad, he figured to grab some education and experience, courtesy of the shipyard, and move on after a couple of years. Once he started as a waterfront welder, though, he never looked back. He became a shipbuilder lifer.

In the Apprentice School, he started welding parts of the *Washington* and working on pieces of the *Stennis* in the shops, and then he finally got a chance to go aboard the *Stennis* in Dry Dock 12. Another thin and wiry steelworker, he got picked for all the tight spots, where he had to twist, turn, and wrench his body into spaces and positions that would challenge an Olympic gymnast. He enjoyed doing that. Still, sometimes, he couldn't get his body just right, so he'd have to use a mirror, which he had to manipulate around one obstacle or another, so he could catch enough of a glimpse of what needed to be welded.

As much as the carrier welding interested him, Carper wanted to do something more. He had always enjoyed drawing, so he set his sights on becoming a designer or working in an engineering department. He attended night school at ITT Tech in Norfolk to learn computer-based designing. When one of the yard engineering departments filled openings in 1996, a couple of years after he started on the waterfront, he made the move to junior designer and then up to senior designer. But he did not like working in a cube farm. He wanted to be part of the real action. He wanted to work with his hands. He engineered another move, to the yard's Waterfront Liaison Group, a hybrid team that worked with designers and those on the waterfront, just as the yard was starting to overhaul the *Nimitz*. The yard tapped him to be a troubleshooter, a fixer, and he relished the role. He used that job to catapult to the position of construction supervisor on the *Enterprise* while the ship was docked in nearby Portsmouth. He took charge of sea valves, making sure they were all replaced and worked to bring enormous volumes of water in and out of the ship. He faced daily battles trying to find enough people, parts, and other resources to get the job done before the dock would be flooded. It was his first year as a construction supervisor, his first year being married, his first time working off-site, and his first time working on the unique ship *Enterprise*. The stress caused a short bout of alopecia; he lost some hair, but he gained insight that would come in handy later.

When he returned to the yard, he found that his superintendent wanted him to take over another important but dirty and challenging task—the painting of the engine room and bilges. Soon after he got the job, he went down to the engine room to survey the scene and assess the challenge. The bilges served as a kind of wastewater collection trough for all the water and gunk that leaked out of the pumps. A lot of cleaning, blasting, and other nasty work lay ahead—just to get everything in shape in the space to be painted. He looked at the general foreman and the painting foreman and said, "Hey guys, I've never worked the bilge. I don't know much about bilges, but I'll learn. I'm going to be relying on you guys very heavily to teach me how you do what you do."

The bilges sat at the very bottom of the engine rooms, below metal walkways and at a higher level than the actual engine machinery. Carper's team removed deck plating to scramble down to clean and blast the tanks and bilges below, kicking up decades of swirling dirt, grime, and rust. Workers replaced parts, prepped the space, and painted in the cramped quarters, setting up tents in the lower levels to protect against paint and debris dropping from work being done on the upper level. Carper managed his own team and also coordinated with other steelworker trades and teams working in or near the engine room. The job required careful choreography.

Management liked his work on the *Nimitz* so much that his bosses drafted him to do the same on the *Washington*, promoting him to construction supervisor two for carrier overhaul. The *Bush* work promised to be different, CVN 77 being a brand-new ship and one that the yard needed to complete and deliver quickly. When he climbed down into the engine room on *Bush*, the scene took him aback. Wooden staging still dominated the space. Chaotic bundles of piping, electrical cables, and other wiring snaked around the compartments. A ton of stuff needed to be finished. Banking on his previous overhaul experience, he quickly formed a plan.

Elsewhere on the *Bush*, Little Ed gained experience as a mechanic on his first carrier, a good gig for a yard pipefitter who had grown up as a son of a shipbuilder, riding aircraft elevators on uncommissioned carriers during family open houses.

He wore his pipefitter's blue hard hat with a jaunty pride. Some folks might think pipefitting relies on brute strength, but it requires much more than just raw muscle. Little Ed lacked the physique of a WWE wrestler, but he exerted enough strength to do the job and, more important, he could picture the trigonometry in his mind for those difficult bends, cuts, and fits. Most people didn't realize how much trig the pipefitters used. They also learned to be metallurgists, to know the different metal types and how they would react in different situations. Little Ed could bend a twenty-foot pipe with six different offsets. He could actually *visualize* the math in his mind. He was using all that calculus he had taken in high school, and he loved it. He had always loved math, even as a kid.

He started working on the *Bush* in January 2008, just after it moved to Pier 2. January on the James tended to be cold. On many a morning, he parked his car in the lot by Thirty-Seventh Street, south of Warwick, at four in the morning and started walking, layered up in his coveralls, thermals, jacket, and insulated boots, to arrive at the gate at 5:15. On some mornings, with that wintry wind blowing from the river so fast and hard, kicking up clouds of dirt and pebbles, he walked backward for blocks to shield his face from the assault during his ten- to fifteen-minute march to the gate.

Once inside, he joined other shipbuilders as they got their daily shift checks and assignments. More than a few gave him a hearty "Hey, Wingnut!" and he responded with a laugh or chuckle. He got his welding line and headed to the metal shack for the metal he needed for the day, and from there he'd make his way to the drawing vault for the ship plans he'd consult during the shift. He loaded up and readied for battle. His work was that of a craftsman.

The secret, as with most things, lay in the fundamentals, in making and double-checking precise and correct measurements. In using the bending blue wire just right, as a visual aid. In getting the trig and geometry just right. In seeing it all in your mind exactly as it should be, based on the sketches, checking it twice and thrice, and then doing it just as you pictured it. In bending it right to match the sketches. In making sure you had the right bending machines. Only a true crafts-man understood.

He knew his pipes the way a professional carpenter understood wood. He checked the size, tossing aside anything over three quarters of an inch if he was going to bend a pipe "by hand." Anything heavier broke the bender—and it would be none too good for the body, either. The yard let them bend only copper when he and the other fitters got to do it out on the ship pierside. Anything else came from the shop.

Bending copper pipe, with its thin walls, really didn't take a lot of strength. You angled it right and aimed the torch at the exact spot to heat and soften the metal, making it as pliable as a pipe cleaner. He packed softer or larger-diameter pipes with sand, so they still held their contours once he started to heat them. He strove for perfection and when he achieved it, it felt like art. A foreman, fellow mechanics, and shipmates stopped and stared and remarked, "Man, that's really some good work." He knew he had done it right when someone from the navy would tell him, "That's some pretty clean work there."

Sure, he could take all this experience, go to Detroit or somewhere like that, and bend tubular frames for cars. He could build all sorts of things. Why the hell would he ever leave here, though? He built the world's biggest warships, for the US Navy. He felt that way even though soon the yard wouldn't let him bend pipes on the ship any-more. It would be all bent before it got to the ship, back in the fabrica-tion shops, for efficiency's sake. The whole idea was to build the best ships, then, now, and in the future. Still—a dying art.

In a Hollywood movie, Bill Bowser would have lived long enough to see Barack Obama elected president in 2008. Life and death usually follow their own course, though, and Bowser's heart finally gave out on him on August 19, 2008. He died at home.

Thus Bowser missed the changes for naval shipbuilding—both in Washington and on the waterfront—that came about after Obama took office. The most obvious of these emerged in and around the Beltway, particularly in the Pentagon, where Obama appointed a seasoned Mississippi politician and former US Navy surface-warfare officer to be the secretary of the navy. Ray Mabus looked and sounded every inch a folksy Southern pol. He had that singsong way of talking to folks, like Bill Clinton, that put even his adversaries at ease. A former Southern governor himself, Mabus had also been Clinton's ambassador to Saudi Arabia. Mabus obviously loved being secretary of the navy. He touted its size and reach, which he planned from the very beginning to expand. His agenda was to increase the size of the fleet. It would never reach the size of the Reagan years during the height of the Cold War, but now the emergence of China provided a greater impetus to push for the bigger naval force. Obama had run on the idea of getting out of Iraq, creating an opportunity to restore some of the naval resources shifted to land forces in the Middle East. The Western Pacific became the focal point for such restoration, to meet the Chinese challenge. Mabus and the rest of the navy needed a policy directing such a shift. The Obama administration put one forward with the Pacific Pivot.

Congressman Randy Forbes had sounded the alarm about Chinese carrier-building plans for a decade. Born and bred in Chesapeake, Virginia, only about a half hour away from Newport News, Forbes believed in his Baptist faith and in the need for a very strong US Navy. When the Obama administration announced its grand pivot in 2009, Forbes emerged as an acknowledged expert on China and a force to be reckoned with. He had become a veteran lawmaker on the House Armed Services and Judiciary committees, as well as serving on the Readiness Subcommittee; the Crime, Terrorism, and Homeland Security Subcommittee; the Constitution, Civil Rights, and Civil Liberties

Subcommittee; and—most important to naval construction—the Seapower and Projection Forces Subcommittee, the last of which he had also chaired.

In 2004, Forbes and Ike Skelton, the ranking member of the Armed Services Committee, traveled to China together. As Forbes and Skelton walked through steel plants, the Virginia congressman, who had spent his entire political career protecting the interests of Newport News Shipbuilding, noticed the steel there—the size, the type, the feel. He knew steel. More than that, he knew navy steel. He looked at Skelton and said, "They are going to build an aircraft carrier."

The older lawmaker shook his head. "You think so? The Pentagon doesn't think so."

Forbes was adamant. "They are. Look at the thickness of steel; look at where they're putting stuff."

When they boarded the plane to fly back to the States, Forbes told his staff, "These guys are serious." And he thought, *Here's the amazing thing—they know everything about all of us. We know nothing about them.*

He returned and perused every committee and subcommittee report or record, House or Senate, searching for anything on China. Nothing. Nobody was raising any issue; no one was holding any hearings.

He would visit his colleagues, who were all glued to TV screens watching images of Afghanistan and Iraq. "Ten, fifteen years from now," he'd say, showing them some image of China, "this is the screen you're going to be watching. You guys are based in the eighties. These curve lines are based in this century, and they're different."

When it came to China and the pivot, Randy Forbes didn't think Obama's people were getting it right at all. They called it a pivot without describing what they meant by that. Still, the pivot helped provide a push for more shipbuilding funding and at least a little cover for the *Ford* development program during the problems with its takeoff and landing systems, its growing price tag, and the busted deadlines. Forbes and a cadre of other lawmakers had managed to keep the carrier hunters at bay, at least through the initial building of CVN 79. It helped to have a powerful local congressman in your corner.

While the Obama election brought about, in an indirect way, a resurgence in the carrier programs in Washington, it prompted another response back in Newport News, at least on the waterfront, tapping a vein of ill will still throbbing after the 1999 strike—the tension between Black and White workers. Because Local 8888 leadership during the strike was largely Black, some of the older White workers—rightly or wrongly—felt that some of the Black workers received more privileges, like access to more strike benefits, than White workers. That feeling threatened to tear apart the fabric of racial unity so carefully and painfully woven by Bill Bowser and Russ Axsom to create the steelworkers force in Newport News. Race relations remained a tricky thing in any city, and Newport News Shipbuilding, a city within a city, suffered the same pangs as any other community. Many wondered how much the racial rifts would affect work at the yard and the cohesion of Local 8888, particularly with crucial carrier contracts coming up.

5

METAL DATA

SPRING 2015

At four in the morning, Little Ed Elliott climbed into his pickup in the predawn blackness and started to plan his workday as he made his way down Route 17 through Gloucester toward York and Newport News. The radio droned on in the background. His mind catalogued, once again, step by step, hour by hour, what he had to get done on this relatively cool and cloudy late spring June day in 2015. The yard expected the navy to finally award the main construction contract to Newport News Shipbuilding for the carrier *John F. Kennedy*, about two years later than initially anticipated.

The yard had stockpiled steel and other components for years, even before the first official cut of *Kennedy* steel in 2011, but the waterfront held off on the *real* work until the yard signed the deal to build the hull, decks, and island and all the associated plumbing, wiring, and systems. Then the yard and navy would discover whether all that digital planning would be enough to meet the cost, schedule, and manpower savings needed for the ship.

Little Ed and the shipbuilders worried about more than the late start. To land the *Kennedy* construction contact, the yard agreed to build the ship in two phases, delivering hull and basic ship systems in 2022 and finishing the combat systems about two years later, instead of delivering the entire ship at one time. The two-stage delivery saved about a billion dollars in the overall cost of the ship, but it also put shipbuilders under the gun to get more of the hull and its systems done sooner than on a normal carrier timeline.

Time to put the pedal to the metal. Ed's job on the *Kennedy* required him to coordinate all the different trades for his spaces. Little Ed couldn't just show up at 6:45, call the foreman, and say, "I need you here or there at seven o'clock." That foreman had already been there for an hour by then and had lined up the paperwork for what needed to be done that day. Little Ed needed to be there before those under him clocked in, put his own day in order, and make sure that that foreman knew exactly what needed to be done at the start of the day.

Making his own way from Mathews County, Lee Murphy also knew that what he did before seven counted as the most important part of that day—the time before he arrived at the yard as well as those precious moments before the start of the shift. Screw up that time and you could forget about the rest of the day, if not the whole week. During those prework moments, he made his calls, setting up one meeting after another, discussing potential problems and what to do about them. You needed all of that under control before the yard shift started. *Imagine what people had to do before they had cell phones.*

The old hands at the yard still struggled with the new way to build the *Kennedy*, with its focus on digital design, engineering, and fabrication. But some, like Little Ed, promoted to a CVN 79 construction supervisor after being a CVN 78 construction foreman, embraced the changes. He walked into the yard in the dewy early morning and out to Dry Dock 12 and the nearby assembly-unit sheds, and he saw almost everything he and his crew needed there already. Pipefitters, machinists, and other trades workers no longer needed to wait to get their pipes, fittings, and other material. The computer reckoned what would be needed and sent that information to the appropriate folks to get the right resources to the right locations at the right time—like having home delivery from Walmart, Target, and Home Depot all at once. It sounded simple, logical, and efficient, but it still was a first on the carrier-building waterfront. Little Ed and his welders stopped by the tool shop and other stations to pick up their pipefitting supplies—filament, torch line, and weld line, for example. But the tablet-toting steelworkers no longer needed to stop by the drawing vault to get those huge bundles of blueprints, even though some older hands still did. Old habits die hard.

Little Ed stopped at his work site and eyed the bundles of pipe sitting by the side of the dock. In the old days, he and his workmates figured which pipes to bend and how. Then they broke their backs all shift long bending metal. Now all that prebent pipe lay at their fingertips.

Computers took some of the guesswork out of deciding what kind of filler material was needed. Welders had guessed wrong in the past. That meant contacting the foreman and then the welding engineers to determine the best fix for the situation. If a pipefitter used the wrong filler material and a welder welded right over it, covering up any imperfections and cracks, the faulty weld joint in the pipe threatened operations and personnel on that US Navy warship. A faulty weld in a pressure system risked a blowout, possibly stranding that vessel at sea or making it impossible to operate safely or lethally, perhaps during combat. It could mean more than just equipment damage; it could mean life or death to thousands of sailors.

The yard used computers in almost every aspect of shipbuilding. The technology offered shipbuilders a whole new world, a sort of James Cameron avatar existence in which workers "entered" a 3D cyberspace hybrid environment of artificial or augmented reality—one being a completely digital animated recreation of a ship deck or ship space and the other being a combination of that digital recreation and the real thing. Workers moved around, welded this, tightened that bolt, ran some wires—whatever they needed to do, digitally, as a dry run to see where problems might be, to discover what might work and what might not. Newport News managers thought the method promised to revolutionize the shipbuilders' training.

Little Ed understood that. The new digital realm helped train the youngsters coming out of the Apprentice School, who faced a harder time visualizing things not actually visible if they weren't in Minecraft or something like that. The company could manipulate this newly created digital dockwork in the same way. But, kibitzing with his father and Lee Murphy during a gossip session outside the gates, Little Ed acknowledged, "I swear I was born in the wrong generation. Like older hands, you get stuck in your ways. You don't want to change." Yet, they needed to change to keep Newport News in competition with the other yards. "I won't say I'm a hundred percent on board, but I'm trying to be," he told the other shipbuilders.

He needed all the help he could get in his new role of coordinating the trades in the *Kennedy*'s spaces. He was like a general contractor who, instead of overseeing one room in the house, now managed different types of workers for multiple rooms in that house. In Little Ed's case, he oversaw the piping that ran through a large area of the ship. He guaranteed the schedule and quality of firefighting, seawater, potable water, chill water, nitrogen, oxygen, and other major piping systems. He also tracked every trades department working in his sections, from the fitters who made the spaces airtight and watertight to the electricians, pipefitters, and machinists who filled the spaces with the equipment and ship arteries that made it all work right. The yard would transfer 2,600 *Kennedy* "spaces" to the navy by delivery. In many ways, those spaces were the building-block cells of a carrier, or any ship, made up mostly of air. Those well-welded empty spaces made up most of the entire ship, aligning with the basic precept of sea-vessel construction. As noted more than a century ago by Assistant Naval Constructor A. W. Carmichael for the United States Shipbuilding Board's pamphlet, *Shipbuilding for Beginners*, "Being made of steel, which is heavier than water, it must be hollow and, therefore, in order to float it must be *watertight*."

Little Ed made sure all those under his charge had all the resources they needed to work according to a schedule free of conflicts and to finish their tasks and jobs in the time and with the craftmanship expected. He scrambled through his spaces, stepping over nests of cabling, ducking under webs of wiring, and navigating around the construction obstacle course that makes up the waterfront scene of a warship being built. He made his way through certain spaces daily, weekly, or even monthly, depending on the type of work and timing of work being done. He scrutinized the quality, where they stood in the build cycle, who was in the space, who was not supposed to be in the space, and who was there who was not supposed to be there. He understood his trade and counted on his foremen for the other trades work. Little Ed acknowledged that the new digital way of building the *Kennedy* made it easier for someone like him to coordinate all of this, to create and modify a plan more efficiently.

With CVN 78 *Ford*, the yard designed and built much of the ship at the same time. It was like baking a new kind of cake by mixing all kinds of ingredients, which you knew *should* taste good together, and hoping for the best. Newport News didn't have a choice when it started the *Ford*, due to the technological advancements and the way the Pentagon decided to develop the new carrier class. But with the *Kennedy*, the CVN 79 construction team started with a full design already completed. With that in hand—or on the computer screen— shipbuilders marked the location of *every* stud on bulkheads with their plasma burners in the fabrication shops before that piece of steel became part of the unit. Workers drilled almost all of the ninety-five thousand bulkhead and steel holes in the shop, an unheard-of accomplishment until the *Kennedy*.

Shipbuilders even treated the cabling for CVN 79 and other Ford-class ships differently. Like the rest of the world, the US Navy designed its modern warships digitally, and Ford-class carriers became the poster ships for that revolution, with a degree of in-ship connectivity and electric power generation that simply eclipsed its Nimitz-class predecessors, whose design dated back to the Vietnam War era. That meant more cabling.

Shipbuilders heaved and hauled more than nine million feet of electrical cabling through the innards of CVN 79, almost twice as much as in the final Nimitz-class carrier, the *Bush*, and enough wire to stretch from Bangor, Maine, to Miami, Florida. The yard analyzed cable installation within assemblies to make sure it outfitted units and superlifts as much as possible, pre-outfitting hangars, shell banks, and wireways. Shipbuilders also developed a software program to help precisely identify the location and route for every cable and every connection point.

Odis Wesby, the new electrical foreman for the emerging CVN 79 power-distribution center, tackled the task of tracking and connecting all that copper wiring—and four million feet of fiber-optical cable—with one hundred fewer electricians than had been doing the job on the *Ford*. He relished the prospect. For the first time, instead of carrying all those cumbersome rolled-up blueprints—like huge cylinders of posters—he and his crew tracked all the equipment, cable lead information, part numbers, wiring location, length, and pathways with their own handheld tablets. They scanned bar codes, and their tablets immediately told them which cable went to which bundle of cables into each bay. They used to eyeball all those wires, hanging down like hundreds of black IV tubes from junctures throughout the compartments, and read the blueprints like those puzzler mazes where you're supposed to find your way out after reaching so many dead ends through trial and error. Now they just tapped their tablets, and the computer did all the work in a sliver of the time it took before. That eliminated almost all rework—a dirty word in shipbuilding. His new, younger workforce handled this all-digital work world like pros as they walked through the yard gates on their first days. Computer savvy, most of them had grown up with some kind of PDA in their hands since their toddling days. They treated it like one big computer game come to life.

Master shipbuilders told those kids, "We'll show you how to use these tools." And those kids retorted, "Don't waste your time. We'll show you."

And they did. Every single day. The new breed of shipbuilders helped the waterfront pros tap the new digital technology, and the old hands shared their tribal knowledge. They formed a bridge across the generations.

To move CVN 79 construction even closer to something more akin to an assembly-line approach, Newport News Shipbuilding developed a "family-of-units" concept, a new approach. With *Ford* and earlier carriers, different units were built in numerous locations in time to support the ship's schedule, like trying to build thousands of cars, one at a time, each from scratch, the total antithesis of an assembly line in a Ford Motors plant. For *Kennedy*, the waterfront arranged assembly units of a similar geometric shape into flow lanes that moved whole units from station to station, where workers performed the same kind of job or task over and over—just like an assembly line—increasing both speed and on-the-job learning among the workers at those stations. In cases where the carrier units were too large to be efficiently moved in an assembly-line fashion, the yard moved shipbuilder work teams from unit to unit to do the same welding, machining, or other hotwork, over and over.

However, try as they might to emulate an assembly-line approach, shipbuilders knew that each carrier was still a handcrafted effort. Each ship took close to a decade to build, with many changes, not only from carrier to carrier, but even from the beginning design to the final one for every single ship. To make it even more challenging, the carrier itself developed a bit like an embryo in the womb of the dock as the steelworkers knitted the ship together. The assemblies shifted on their foundations, and the metal "breathed"—contracting and expanding with the cold and heat—and moved to follow the sun the same way a plant did. No two carriers came together in the same way.

While Little Ed focused on the *Kennedy*, other shipbuilders and the navy recalibrated the image of his former vessel, the *Ford*. Yard executives started to see a sea change in the faces and the comments of VIPs from Washington and the Pentagon who came to see the progress on the *Ford* firsthand. All those lawmakers and Congressional aides made that drive down to Tidewater with preconceived notions of the *Ford* and walked away with a different point of view. Tours became so common that steelworkers paid little or no attention to the intrusions. None of the visitors had really considered the innovations on that ship; they simply saw more dollar signs. But they came away amazed by the technology and the progress, a huge improvement over the *Nimitz*. The level of detail stunned them. They began to see the *Ford* as *the* ship to take on China, *the* ship for an increasingly dangerous world.

The yard persuaded most that the *Kennedy* promised even more improvement. The company wanted to put that into writing, to make it official. With the drafting of the official contract to build the ship, the navy and the yard agreed to take a huge risk together. The yard committed itself to fix all the mistakes from CVN 78 with about a fifth less manpower to build CVN 79, even as the navy was still putting *Ford* through its paces to find out what problems remained and the extent of them.

The navy's stick if the yard failed? The end of carrier-building schedules for the foreseeable future. The possible carrot? A two-carrier deal for the next ships, which would not only would mean work for years to come, but also the chance to again prove that Newport News could build carriers efficiently and affordably, as with previous multi-carrier contracts. In that scenario, the yard figured to guarantee work for decades. As the former yard president and current CEO of the yard's parent company, Huntington Ingalls Industries, Mike Petters knew how important such multi-carrier deals could be to the yard. When he ran the carrier program back in the 1980s, the yard was on the back end of back-to-back two-ship buys: the *Abraham Lincoln* (CVN 72) and the *George Washington* (CVN 73), and the *John C. Stennis* (CVN 74) and the *Harry S Truman* (CVN 75), as close to a production-line feel for total nuclear carrier construction work as the yard had ever seen. Barely had one carrier been launched out of Dry Dock 12 before another one was started in the same space, with the same workers, the same tools, and the same basic resources already in place, putting into practice the institutional knowledge and lessons learned straight-away, meaning faster, cheaper, and better work.

After *Stennis*, they had started assembling the metal in the dry dock for *Truman* in about a month. But the two-contract deals ended, and after *Truman*, it took eighteen months to start on *Reagan*. After *Reagan*, it took about eighteen months to start on *Bush*. Following *Bush*, it took twenty-six months to start on *Ford*. The nation lost a whole carrier's worth of work in those missing months between ships. Now, with the new digital waterfront being established for *Kennedy*, the yard hoped to make up that time, with shipyard workers viewing vi-sual building sequences the way others would access information via YouTube videos.

It would be hard for a non-shipbuilder, even a non-carrier ship-builder, to appreciate the technological sophistication achieved for the *Kennedy*, that is, to digitally pinpoint the location of all those studs and holes at any given moment during the construction pro-cess, because, as noted above, the *Kennedy*, like other carriers under construction, was a living, breathing thing, moving with the sun, heat, and cold. All piping, wiring, and millions of tiny pieces moved, too. Daily shipbuilding was like performing an operation on a patient who was awake, fidgeting, and knocking out of sync every preplanned scalpel cut or closing stitch. Through the decades, the fitters, welders, and other crafts workers calculated and recalculated every day and then throughout the day, using rulers, pencils, and some common shipbuilder sense. Before *Kennedy*, steelworkers waited until sections or compartments were already aboard before performing some hand measurements, bulkhead cuts, and so on.

The ship was pockmarked with 350,000 electrical studs, plus well over a million for insulation, all of which workers had located and installed by hand on previous carriers—hundreds of shipbuilders climbing up and down ladders, rulers and chalklines in hand, marking stud spots and installing them, each three to four inches long, using brute strength to get them in, over and over again, hour after hour, day after day. Wiring racks and cabling racks—they studded everything that needed to hang, including all the pipe hangers, thousands of them, all needing studs every few inches, in every space, throughout all the thousands of spaces. On the *Kennedy*, the plasma burner did most of that.

The digital work didn't eliminate shipbuilders, but it made their jobs easier. Using bar codes and tablets, electrical foreman Wesby and his team retrieved all the information they needed about cabling and connections. What's more, immediate access to all this digital information ensured the correct cabling length, connections, and other important attributes. Before, the work involved a degree of trial and error and no small amount of rework. Now fewer, and in many cases younger, computer-savvy electricians performed the task faster and more accurately.

With *Kennedy* construction ratcheting up and the navy keen to buy even more nuclear-powered ships, the yard began to augment its waterfront force. Normally, it took years to train new crafts workers to the required proficiency, but yard managers reckoned to shorten that whole process. The yard tapped the new generation's digital savviness to help bridge the gap between experienced and inexperienced shipbuilders, using colorized digital models that users could rotate and orient according to whatever was being worked on, with augmented reality, so they could see what the complete space should look like. The younger workforce intuitively got it and learned very quickly, even teaching the older hands how to build their ships better with these new tricks. At the same time, the seasoned shipbuilders acted like waterfront Yodas, imparting their knowledge, but thanks to the new digital visualizations, at remarkable speed. By waterfront custom, yard supervisors set two craftsmen to build the same compartment, one a seasoned journeyman, the other a rookie. Using augmented reality, the newcomer generally finished the space in the same amount of time as the pro, flattening the learning curve.

Little Ed also took notice of the computer whiz kids. Chronologically, he was a pseudo-member of that generation, but he remained old-school, more likely to carry a clipboard than a tablet. Instead of opening a laptop, he preferred to pull out the actual drawings, spread them out, and study them. He avoided computers. When he sat at one, hunting and pecking away at the keyboard with a fresh handful of wingnuts strewn over his desk, he thought, *I'm a shipbuilder. I'm a doer.* He wanted to get out on that ship and drill holes in steel. Little Ed knew, though, that he needed to learn, to adapt, just as the yard every so often had to relearn its business.

Lee Murphy, too, eyed the new kids on the yard, noting that many looked older than he'd been when he first entered the gates. The success of the *JFK* would depend on these youngsters, and they looked to be up to the challenge. "Smart as a whip," he said more than once to another seasoned shipbuilder. They all voiced good ideas. He considered himself smart, but they seemed smarter; they'd grown up with iPads and iPhones. He, on the other hand, remained computer- and iPhone-challenged, but because of his age, he told other shipbuilders, "I just don't care. I'm as dumb as clams when it comes to all this computer stuff."

When he came into the yard, he learned about shipbuilding from other craftsmen, from foremen—specialists. Master shipbuilders. Or, as he put it, "Old farts like me." To their credit, though, these young people realized they didn't know it all. And the yard did a good job of pairing young digital Jedis with old farts.

Hurdles remained, though. A barrier still existed between the outfitting and construction sides of the house. Murphy would climb into an assembly unit, ready to install it on the ship, and just by eyeballing it, he could see they had put the piping too close to the place where they needed to join it to the other erected units. The people planning it with all these computers lacked input from seasoned shipbuilders' common sense, and now, as he tried to put two units together, the pipe for one was running six inches from where they needed to fit and weld. He eyeballed the misalignment and thought, *That's not good.* Still, Mike Butler and the yard wanted to build the ship with these tools. They wanted to rethink everything. To use a baseball analogy, they were "rounding third and heading for home each time"—knowing full well they'd get thrown out once in a while—but they'd also wrack up the runs.

They did not expect the digital waterfront to form suddenly and magically. A computer could never know the feel of steel, the different kinds of metal needed for different parts of the ship, to make sure that steel would flex when it was supposed to and hold firm when it needed to. The erector steel used for CVN 78 *Ford* served as a cautionary tale and a case in point. The yard relied on computer models to determine where to use the thinner steel on parts of the *Ford* where it would cut cost and weight. In the end, though, steelworkers wound up doing a lot of deck straightening—cutting deck pieces and straightening them all out. The yard learned its lesson and used thicker steel for platforms and decks on the *Kennedy*.

Those learned lessons encouraged Butler, who told everyone to think *way* outside the box, starting with questions like: Can we build a superlift from the dock all the way to the flight deck? One huge ten-story superlift? Sure, they faced limitations, like the laws of gravity and physics, but they strove for a plan to build a carrier in much the same was as you might build a skyscraper. *Kennedy* really was, in a way, a seagoing skyscraper. That idea changed the whole way they viewed shipbuilding. They never achieved that one gigantic superlift, but they did assemble giant superlifts. They never created a superlift from the dock floor to the flight deck, but they achieved one from the dock floor to the main deck and then to the flight deck.

Look at the Brits: they built a carrier in five big pieces, in different yards, and rolled them all together. And even back in the 1970s and 1980s, when the yard built jumbo tankers, the steelworkers took a five-hundred-foot tanker, cut the bow off, cut the stern off, built and inserted a six-hundred-foot section in the middle, and welded the three pieces together. They needed to think about shipbuilding from a different point of view. Instead of building a ship along horizontal lines, like layers of a cake, the yard would have to think about building the vessel in vertical lines, like slices of a cake.

Kennedy shipbuilders talked to skyscraper builders in New York and asked them, "How do you build an eighty-story building? What if that building was not anchored to solid bedrock? If you had to move that building as you built it—how would you do that?" By thinking that way, steelworkers did things like combining nineteen lifts into one superlift and fourteen more into another one.

Every one of those bigger superlifts saved time and money, making it possible to build *Kennedy* cheaper and faster. Butler encouraged his team to think beyond just building bigger superlifts. He wanted to know: Could they design the ship differently? How could they order materials differently? How could they cut the steel differently? How could they plan the whole thing differently? While Butler started with an overall full-ship design in hand, *Ford*'s operational successes and failures had already prompted the CVN 79 team to make design or construction changes, even with major systems like launching, recovery, and weapons elevators.

Butler and his team opened the aperture for new ideas in way the yard had not done before for a carrier. A lot of little things, Butler liked to remind folks, added up to big things. Becoming a better hitter in baseball, he would say, could start with changing your stance in the batter's box. Babe Ruth had one of the ugliest stances in the game, yet he hit more home runs than anyone in that era. How do we all perform like Babe Ruth but do it better? He pushed his people to imagine themselves anointed king for a day. "What would you do? What would you change?"

The lack of an official contract for the *Kennedy* hamstrung the shipbuilders about what kinds of changes they might make. Due to all the issues with *Ford* and the military funding concerns in the Beltway about carriers in general, the *Kennedy* contract faced delay after delay. When Mike Butler got the *JFK* job, the yard had yet to determine the official cost estimate and schedule. To keep the ball rolling, the navy awarded a series of small contracts to buy long-lead-time material and start some of the early manufacturing work on *Kennedy*. Thus the yard started to buy some of the castings, steel, and propulsion equipment, the kinds of things that can take a long time to get once they've been ordered. Butler and his team needed to make sure they had those big pieces and steel slabs on hand when they started that early construction. With the *Kennedy*, Butler wanted to get a good start when his crews began actual assembly. More early components included nuclear construction material, some of which the government furnished, some which the yard fabricated, although shipbuilders kept the exact nature of the nuclear shipbuilding work close to the vest, due to its classified nature. The real shipbuilding truly began, though, when the yard got the full construction contract. Everything else was constructional foreplay.

Once the yard secured that deal, Butler planned to focus on building the assemblies, laying the keel, and the erect plan—the contract would specify the devilish details. *Kennedy* workers kept busy without the signed deal. As the long-lead-time items came in, as the yard started to get the steel it needed for the ship, and as the fab shop started to turn out some CVN 79 parts, the waterfront was humming with *Kennedy* work. Ship bits, units, and pieces started to pile up, scattered in and near the dry dock.

Butler walked around and checked the various units in various stages of assembly scattered about, like a bunch of mini–construction sites instead of a single one. The dock resembled a demolition zone. Looks deceived, however; the dock operated like a well-tuned machine. Certain cars, for example, look like wrecks on the outside, but what truly matters is the car's engine under the hood.

As boss, Butler liked to check under the hood. He wanted to see the motor. He made his rounds, stopping to ask the foremen, the welders, the painters, "What's happening? How's it happening? What are you doing? Who's doing it and why are you doing it?" A pulse check? Sure. But he also wanted to know because he wanted to *learn*. Every day.

Butler strove to scrutinize every step of the process, but he refrained from interjecting himself at every step. He checked to see if there were any barriers—missing paperwork, an undelivered part, that kind of thing—that he, in his position, could break loose. Or maybe somehow he could help his people get the *JFK* work done more efficiently. He could never micromanage every department, nor would he want or need to. He watched foremen, riggers, and mechanics move steel and equipment around the dock and felt reassured: these subject-matter experts knew their crafts and arts the way members of an orchestra know their instruments. He served only as the conductor, directing them so as to play their best, helping them get their instruments ready and their skills in place. His style was to let them do their jobs, let them manage their own business, not stepping in unless it was truly necessary. In a ten-year, carrier-building marathon, opportunities would arise to do that, but you had to pace yourself, sprinting only when you needed to.

The sprinting for the carrier *Kennedy* started after a cloudy and cool but muggy late spring day in June 2015, when the shipyard got the official main contract for just over $3 billion to build CVN 79, with an additional $941 million thrown in to sweeten the deal—a total of more than $4 billion. The contract called for Newport News to build the ship, obligating the navy to furnish the vessel's main systems, such as radars, reactors, and aircraft launching and recovery, which account for more than a third of the total carrier cost. With that contract award, shipbuilders truly started to polish the design, work out the engineering, buy the bulk of material and hardware, line up the logistics, and begin the real construction of the *Kennedy*. Thousands of steel, pipe, cable, paint, and equipment suppliers around the country prepared for the orders that kept many of them in business.

No waterfront celebration marked the contract milestone. Shipbuilders surveyed the dock and surrounding area, mindful that 450 of the ship's 1,100 structural units had already been built and awaited the lifts to mate them to the carrier hull. Now, workers geared everything toward that first superlift of assembled and outfitted units as the steelworkers prepared for the ship's keel laying near the end of the summer. Thanks to digital engineering, shipbuilders moved bigger and more complete superlifts than they had ever attempted before.

During that steamy July following the CVN 79 contract award, both the steelworkers and the managers on the Tidewater waterfront felt the heat to ready the bits and pieces, as well as Dry Dock 12, to stay on schedule for the first superlifts and the keel laying. After troubleshooting the engine-room spaces for the *Bush* and the *Ford*, Sam Carper found himself drafted to work on *Kennedy* in a much bigger role—moving all the ship parts through the final-assembly platen to build the units that form the superlifts for the ship, passing them all on to Lee Murphy in the dry dock.

Until CVN 79, the platen had served as a mini–Steel City, focused on grinding, welding, and other steel-heavy trades work. But digital engineering and improved build strategies changed that by moving more of the outfitting and the trades work upstream, back into the shops and off of the platen. Like many others, Carper became a true believer. Instead of seeing small pieces of the ship on the platen, he saw whole sections that actually looked like slices of a ship.

To better take advantage of this new digital shipbuilding success, the yard also erected a new Unit-Outfitting Hall. Located on the waterfront just a bit north of Dry Dock 12, the three-bay building resembled a warehouse or a small factory. When Carper's teams started outfitting and mating sections on the platen, steelworkers would start to use the unfinished building even as craftsmen finished putting it together. Carper often ducked inside—grateful, like other workers under its cover—to be shaded from the brutal Tidewater summer sun.

Carper, though, couldn't escape the heat of quickly approaching deadlines that he might miss because he lacked the right supplies. Just as his craftsmen started to piece together CVN 79, they found a problem: the vender had made all the pipe fittings too thin. Carper halted piping work on the platen, and the shops stopped using those fittings while the vendor remanufactured and resent a new batch with the proper thickness. The steelworkers made sure no improper fittings made it into the units. A bad piping system put the whole carrier and its crew at risk.

When the new fittings arrived, Carper had another problem. To finish the superlift on time, his craftsman needed now to do all the work associated with those pipes, including the outfitting, while other steelworkers were doing all the heavy structural work. As welders mated the long seams of the overall superlift in a smoky shower of sparks from carbon arcing and other metalworkers ground away at the rough steel surfaces, fitters installed pipes and associated components, trying to keep the material and their equipment from being damaged amid the fireworks engulfing them. Carper, who often had to manage thirty to forty units being built all at the same time, became accustomed to dealing with issues like bad fittings.

The first steel moved for the *Kennedy* keel came to the platen as flat slabs, some twenty by thirty feet square, which steelworkers joined to form the width of the ship's bottom. Shipbuilders called them egg crates—like big square metal mazes, welded to the tank section. The pipefitters scrambled through the mazes, installing the pipes in those steel puzzles and inside the tanks. Welders sealed it up. Riggers hooked the heavy crane chains to the sides of the pieces. A siren sounded. Big Blue flipped the section like a huge metal pancake—quite a sight to see as Big Blue lifted panels of steel the size of a dozen billboards off the ground by a handful of cables and flipped them end over end like some square metal acrobat, before setting them back down on the ground, while steelworkers blew whistles in short staccato bursts to alert anyone in hearing range that something dangerous hung overhead. With all the pieces and tank top welded together, the assembly resembled a massive plate-steel sandwich.

Rising 235 feet above the waterfront, Big Blue was the biggest gantry crane in the hemisphere. It straddled Dry Dock 12 like a sentinel, steel legs planted on opposite edges of the 549-foot-long concrete dock trough. Emblazoned across the top of the five-thousand-ton crane: NEWPORT NEWS SHIPBUILDING. Some folks driving by on the highway mistook it for the city's welcome sign. The view from the top girder, reserved for seagulls and the precious few steelworkers with business there, provided the best perspective on the scope of the yard—550 acres of sheds, cranes, dry docks, piers, a firehouse, and brick buildings—a construction plant sprawled across the shores of the river banks along the mouth of Chesapeake Bay, the nation's own inland sea. It offered a breathtaking view of the spot where the *Monitor* and *Merrimack* fought the world's first ironclad battle during the Civil War and revolutionized the design of warships.

When assemblies made it to Dry Dock 12, they came under the care of Don Doverspike, a key *Kennedy* construction supervisor who worked for Lee Murphy. They both knew they had each other's back.

After more than a decade of doing such work on carriers and about three decades of being a Newport News shipbuilder, Doverspike knew his business. The yard had drafted the seasoned troubleshooter to supervise the troubled *Bush*. Carrier officials counted on his steel spine to put and keep *Kennedy* on track. ·

The course of steel ran deep in Doverspike. The come-here was born and raised in the western Pennsylvania town of Aliquippa, just northwest of Pittsburgh on the Ohio River in Beaver County. In the heart of steel country, the town served as the cauldron for the plates and beams that formed the Tidewater-built carriers. When Doverspike made his way to Virginia in the mid-1980s, about eighteen thousand called Aliquippa home. By the time the yard prepared to lay the *Kennedy* keel in the summer of 2015, the population had been cut in half.

Like many strapping lads, Doverspike forged his youth in an Aliquippa steel shop, before the industry started to melt down. Then he took his skills down to Tidewater, starting as a yard X-11 shipfitter in 1985, installing carrier hatches and doors. He became an expert, plying his trade on the carrier *Theodore Roosevelt* and then the *Lincoln*. After that, he went to the *Washington*, the first ship he worked from beginning to end. After running the final assembly platen for the *Ford*, Doverspike had a fairly good idea how to prepare Dry Dock 12 for those carrier superlifts.

First, steelworkers cleared the dry dock of any material not needed to build the *Kennedy*, then installed, built, and created all the infrastructure the steelworkers would need to erect a nuclear-powered carrier. They hooked up temporary gas for welding and burning, air compressors for steelworkers to run their drills, wrenches, and other tools, and the fans and ventilation systems to remove smoke and clear the air in all those spaces, units, and compartments that would be filling the dry dock. Lines, hoses, and wiring snaked everywhere. The constant whir of machinery compressing air, generating electricity, and ventilating the stale and sometimes deadly air filled the waterfront as needed for the next four years.

Doverspike worried about more than just the initial dry-dock prepping. Remember, the yard built carriers from the bottom up and from the middle out. After each superlift and section of ship, Doverspike's team prepped the next area, moving all those wires, hoses, and other equipment to the next section of ship. The steelworkers built the ship upward, standing on each deck's finished spaces to build the deck above. When it was time to move a level higher, Doverspike's people moved all the equipment and construction material inboard and set up and prepped for the next piece. Once that got done, they moved everything back on top and started building from the middle again. They did that for every level. And the steelworkers never worked on just one level at a time. Every time, they repeated the process and did it quickly.

Oddly, the first superlifts for *Kennedy* became the easiest, because Doverspike worked from a clean slate. Doing so became harder and harder further along in construction as steelworkers built up level by level, out toward the aft and the bow, setting up multiple pieces in a week. Whether in the very beginning, in the middle, or near the end, Doverspike maintained the same mind-set about the job. Never a desk jockey glued to his office, he made his rounds, walking the waterfront, climbing up and down the scaffolding and in and out of the dry dock, asking what was going on—face-to-face and checking the work with his own eyes.

"OK, this is what's coming next. What are our next actions?" he'd ask.

He inspected. Then reinspected—and reinspected again. Of course, he checked the work with the drawings. He also employed other tricks in the dark shipbuilding environment, in all those dark spaces. The craftsmen all knew about Doverspike's flashlight. He'd enter a space and lay his pen flashlight flat on the ground.

"You'd be surprised what you'll find," he told the craftsmen as he read the shadows his small light created. Things popped right out, like a distortion in the metal caused by welding, showing up like some misshapen creature, or some studs not cut off a bulkhead properly, a common-enough problem. He checked everything in the compartment, his seasoned eye catching other common mistakes, like mislabeled pipes or overhead electrical cables missing those tags showing where they were meant to run to. No computer matched his uncanny instincts. Sometimes, the welding simply failed to meet shipyard and steelworker standards. He'd check another compartment and find metal burned by welding and some welds with holes in them, and possibly impurities in the gaps. He'd lay down his flashlight and reveal signs of spatter—little pieces of welding that had rolled off. He'd put his flashlight on the top of the compartment and find some areas that weren't quite mated correctly because of some welding mistakes.

No one expected zero welding mistakes during the building and erecting of an aircraft carrier. Given the amount of welding on the massive warships, steelworkers made fewer mistakes than some might expect. Foremen often gave welders a certain amount of leeway in terms of the quality of their work. In some cases, Doverspike or one of the foremen for the job could bring in a special inspector to check work they had issues with. But it very rarely came to that. Most welding passed the Doverspike flashlight test because the welders, foremen, or even other craftsmen carefully checked the work. Some of this all came with experience, of course. Younger guys learned how to do those difficult vertical and overhead welds, but until then, the seasoned shipbuilders helped them along. They all worked as members of the same team, and they didn't want any questions about whether their work met the quality tolerances. The foremen pushed for the quality, and they knew which workers delivered and which ones did not. Still, when Doverspike came by with his flashlight, they wanted all the work to shine.

Unlike some obsessed traditionalists, Doverspike swore no allegiance to old ways, no matter what. He embraced any kind of new technology or mind-set that helped him and his team get the job done faster and more efficiently. On the *Kennedy*, for example, they started a pilot program using digital work instructions on tablets. Down in one corner of his office on the platen, his team built side shell units—

mating two fabricated steel pieces tighter with all the piping in place for the shipfitters to set into tanks. The craftsmen relied on drawings, or visual work instructions, for a step-by-step description of how to do the job right—how to run the pipe, mark the proper measurements, and so on. Some likened it to IKEA, but with a lot more detail that made sense. For this particular job, the yard also put those visual work instructions on tablets.

The craftsmen started using the tablets and, after a few false starts with the digital equipment (anyone trying some new phone or laptop app knows what that's like), they joined the pieces with the video and timer running. Then the team did another job using the tried-and-true old-fashioned drawings—and did it faster than they had with the tablets. Doverspike laughed heartily watching it all play out. He figured the team did better on the drawings partly because of their inexperience using tablets and partly because they already had the experience of doing it the first time with the tablets before trying to use the drawings.

Doverspike found nothing fun or funny about one particular type of company transformation—the move toward greater safety. When Doverspike patrolled the platen or the dry dock, he delegated some of the quality-inspection burden to his foreman. But let some welder, shipfitter, or craftsman forget a hard hat, lean up against a safety railing, or commit some other safety infraction and he turned into a drill instructor who just found some recruit drunk in the ranks. Safety emerged as the one thing that had come a long way since he'd started at the yard. He recalled the days they didn't even have ventilation to suck away deadly welding fumes. Now they knew better. And yet, the waterfront workers continued as their own worst enemies.

Roaming the construction zones, he scanned the craftsmen to make sure they were wearing their personal protection equipment, their PPE, *perfectly*. Instead of looking at faces, he watched for safety glasses. Instead of scanning for badges, he scrutinized respirators. More than a few steelworkers quickly checked themselves to make sure they were squared away with PPE; they knew what it was like to get a Doverspike ass-chewing. They noted his obsession and they knowingly smiled when he approached someone leaning the wrong way at the wrong time on a safety rail.

"You're leaning on that safety rail," Doverspike would yell, emphasizing his point with hand gesticulations. "Did you weld that? Do you know how good that weld is?" He wouldn't wait for an answer. He didn't care about the answer. He already knew the answer, and it was not in the violator's favor. "Don't lean on that unless you're the guy who made sure it's not going to slip! The number-one fatality in shipyards are falls. That's a sixty-foot drop! On concrete!"

He kept after them, all the time. Especially the young ones. "You're young. You're carefree. You don't think you need that safety belt." But he let them know loud and long that they most certainly did. He got quite animated. Red-faced, finger pointing, and on point.

Sometimes he tried a different tack. "I understand that it's not comfortable wearing that hard hat, especially with some of the positions you need to get into when you're in these compartments. But then you start moving around this compartment and you hit your head and my old butt has to carry you out here because you knocked yourself out.

"And I know you don't feel you need those safety glasses on, but then somebody moves the ventilation, and you get something in your eye. When you're finishing out this compartment, it's not like your house. If you go up on that ladder, you're liable to hit your head on that hanger—so you need to put your hat on."

He'd finish with a warning: "If you don't do it right, people lose their lives. It's a dangerous place to work."

His mind focused straightaway on safety when he woke up before the sunrise so he could leave his home near Lightfoot and Croaker in time to be sitting at his desk before six. He lived only about fifteen to twenty minutes away with no tunnel or bridge traffic to fight through, but he felt compelled to be ready to address any hurdle before seven.

While problems and issues ran the gamut, they generally came down to trying to stay on budget and on time. It drove him crazy to miss a deadline and hold up somebody else's work. Still, they always seemed to want more, faster. He had a saying—punctuated with that normal hearty laugh—"It always seems that you're running toward the finish line and finish line is always moving."

He found that particularly true on *Kennedy*, where the pressure was building. Better, cheaper, faster. With quality and on deadline.

August 2015 started off with a US presidential debate the likes of which Americans had never seen before, with Donald J. Trump turning politics on its head in the first televised battle among the Republican candidates. While Trump loudly berated other GOP stalwarts and the Democrats for creating what he described as an American mess, the country was still showing signs of digging out of the worst economic slump it had seen since the Great Depression, reporting a seven-year low in the unemployment rate. Around the world, people in other lands still viewed Americans as saviors. On August 22, for example, three Americans and a Brit overpowered a man armed with an AK-47, a pistol, and a box cutter as he walked down the aisle on a train outside of Paris. Back in Newport News on that very day, as a muggy morning of clouds turned into a beautiful and uncharacteristically dry sunny day, on the Tidewater waterfront, steelworkers started some truly substantial work on the carrier *Kennedy*.

On this clear and comfortable Virginia summer morning, CVN 79 reached one of the biggest construction milestones in the carrier's young life. That day, the shipbuilders laid the ship's keel and officially started to *erect* the ship.

Most of the shipbuilders just loved this moment. Sure, VIPs flooded into the yard, followed by the media horde—the public acknowledgment of it all. But the *tradition* of it all mattered more than any of that for those on the waterfront. It harked back centuries, back beyond the time of the Phoenicians. In the days of wooden boats, they actually laid down a center keel—the ship's spine and backbone. Newport News shipbuilders did carriers quite a bit differently now, starting first with the pomp and circumstance. Every US Navy warship had a little of that, but a carrier had it in spades. And a carrier to be christened the *John F. Kennedy*—the second flattop to carry the name—attracted more brass and bright lights than just about any other.

That attention started with the ship's sponsor, the person most officially connected in the public's eye with the vessel from the moment it was named through the rest of its life. For CVN 79, that person was Caroline Kennedy, the daughter of JFK and the first woman to be appointed ambassador to Japan, by Barack Obama. She was also an attorney and a bestselling author of books on the Constitution, civics, and poetry. She also had christened the first, non-nuclear carrier named the USS *John F. Kennedy* (CV 67) in 1967. Kennedys and carriers just seemed to go together, and her name still retained a kind of magic along the waterfront and in Newport News. Unfortunately, Ambassador Kennedy could not leave her post to make the keel-laying, but she did send her signed initials. Steelworker Leon Walston stood on the dais with honored guests, donned the white-and-gray protective hood and the bulky dark yellow gloves, grabbed his torch, and amid a small shower of fireworks and sparks, welded her initials onto a steel plate that would be permanently affixed to the ship from that day forth. The others on the stage also donned the protective headgear, making them look like a squad of pseudo–Star Wars stormtroopers.

A Kennedy did come for the keel-laying—US congressman Joseph P. Kennedy III spoke at the event. He also toured the yard, stopped to meet the waterfront shipbuilders, shook their hands, and smiled for pictures, one of which he hung on his office wall from then on. Considering the heritage of his last name, one might assume the Massachusetts congressman might feel nonchalant about the whole naming-the-carrier-after-his-famous-relative thing, but he and his whole family felt honored, and he personally felt humbled by it all. When he saw the first sections of the ship in that massive dry dock, his jaw literally dropped at the size and scope, the feats of engineering that went into this giant warship. He could not wrap his mind around it all, the marvel of the vision, the genius of the engineering, and the sheer determination of the workers.

He watched the cranes pick up slabs of metal weighing hundreds of tons as though they were squares of plastic and fit them into place with the precision of a surgeon while shipbuilders welded it all together, a mating of metal that would last for half a century carrying thousands of sailors and tons of weapons on the world's savage and salty oceans—not a recipe for steel longevity. The carrier was the number-one American-labor-made thing that proved the old maxim of the whole being greater than the sum of its parts.

This ship would carry the name *Kennedy* wherever it went. It embodied the efforts of thousands of shipbuilders over a decade, with the support of the entire Newport News community, which the congressman found stunning. CVN 79 *John F. Kennedy* also would carry thousands of sailors, from all walks of life, all different backgrounds, who would serve in a sense as ambassadors for the US, not only showing the military muscle of the American navy, but also to show American compassion with humanitarian aid in times of crisis or disaster. In his travels as a congressman, Joe Kennedy was struck by the number of people in other countries in, say, the Western Pacific, who would approach him directly and tell him of the happiness and hope they felt when they saw an American carrier on the horizon, after a typhoon, ready to provide the aid that only that US Navy ship could.

As part of this visit to Newport News, the congressman also toured the *Ford* as the shipbuilders were putting some of the finishing touches on it, and he felt assured that it would meet US military needs against China and other future foes. He also believed that the *Ford*, like carriers before it, would anchor US naval missions to guarantee open sea-trade lanes around the world. Many people failed to realize the importance of those operations not only for direct American commerce, but also for trade between American partners and allies, which, in the end, furthered American interests. Carriers made it possible for the US to set the global trade table—and for folks to go to a Walmart for affordable shopping with an ease most of the world envied. The US reaped most of the benefits of being the global maritime cop—and the best way to flex that muscle was a carrier and its strike group on patrol within striking distance of some distant shoreline, keeping the peace through a show of strength. China might be buying more carriers to boost its maritime might in the Western Pacific, but it could never extend its carrier operations farther, unable to match the reach of the US force. The American carrier gave the US a unique strategic advantage. Full stop.

Another keel-laying speaker was retired rear admiral Earl "Buddy" Yates, the first commanding officer for the initial carrier *Kennedy*, wearing a new royal blue *John F. Kennedy* cap. Lead Rigger Mike Williams handed his walkie-talkie to Admiral Yates, who clicked it to transmit the order to Big Blue: "Hoist this keel unit into the dry dock."

Hundreds of spectators lined the edge of the dry dock, looking on with phone cameras at hand as Big Blue's alarm blared and the crane wheeled into place in a cacophony of whirs, clangs, and squeaks.

The yard entrusted the operations of that mighty machinery to Charlie Holloway, a master shipbuilder with than four decades of experience at the shipyard. His hearty, infectious laugh ingratiated him to just about everyone. Of course, when your job meant sitting more than twenty stories up in the air in a giant glass-and-steel cross between a diving bell and a lunar module, you didn't get to share too many laughs with other steelworkers. Holloway was one of the few working directly on the waterfront (over it, to be more exact) who could claim a paycheck to sit in a chair all day—and a darn comfortable one at that. Leatherbound cushions with the best swivel action and a state-of-the art joystick with computer screens mounted on both arm sides—everything a steelworker needed to operate the most powerful shipbuilding gantry crane in this half of the globe.

Sporting a head of salt-and-pepper hair, Holloway made a daily climb to swing his stocky frame into his perch. While he took an elevator—about the size of a small closet—most of the way to his chair, he then had to pull up a trapdoor and climb down a forty-foot section of metal stairs to his bubble—temperature-controlled to keep him cool during the hottest days of summer and warm on the coldest in winter. Wind, though, could get his "office" swaying, and folks had been known to get kind of seasick visiting his workspace.

Staring out of steel-framed rectangular windows, which curved as they tapered toward the floor, he literally had a bird's-eye view of the NEWPORT NEWS SHIPBUILDING main gantry bar that gave Big Blue its name, as well as of the dry dock below, now looking like an empty cement swimming pool, and of the total expanse of the waterfront. The James rolled by like a blue ribbon of steel, and the bridge spanned it. The shops, the steel fields, smaller cranes—it all seemed chaotic down there, but up here, it all seemed to make sense. Hard-hatted steelworkers swarmed about everywhere, the white hats farther away and the orange ones marking the riggers beneath the huge crane. Holloway saw them all, out his windows or through his glass-bottomed, steel-framed floor.

The crane operator had started as a rigger, and he knew that communication with the ground crew via walkie-talkie was vital for safe and successful Big Blue operations. He trusted his ground crews and his own craftmanship, so lifts never made him nervous, and he feared no height, one of the most important job requirements. He had backed into the job, just as he had backed into working in the yard at all.

A born-and-bred Newport News lad, he did indeed yearn to be a builder—of houses. Particularly his own. He wanted to lay brick and build his own brick house. He started out on that quest after graduating high school in 1978, but when the local construction market collapsed in 1980, he got a job at the yard, strapped on his tools, and walked through the gates. He liked his job but wanted a shot at running the crane. When one of the crane operators failed to come to work, his supervisor gave Holloway a shot. That was a few years ago, and he had been climbing up to this seat every day since, after walking to the yard from his home right up the block on Forty-Ninth Street.

He settled comfortably in his chair; he could spend whole eight hour shifts there. He leaned over to get a better view, his hand loose on the joystick to vertically control his hoists—two would be attached to the front and one for the rear, which made it possible to lift, lower, trolley, and, as shown before, flip. He could lower just one side and raise another or hold one side steady and trolley the other into position. His job was to precisely dangle a small mountain of steel, between 400 and 1,050 tons, in space.

The crane's chains descended like steel vines, and riggers scrambled to hook them just right on the edges of the ship piece, the engine room number-two unit, which looked like the cutaway of an entire smaller ship, with hatchways, bulkheads, and large spaces for machinery. With its well-weathered green-and-rust surfaces, the unit already looked as if it had been through a war. The lines having been hooked, tested and cleared, Big Blue started to lift the piece with a measured, maddening slowness, as steelworkers blew whistles, waved arms, and directed the move as best they could.

On the side of Dry Dock 12, steelworkers and other spectators followed every inch of the move—a helluva thing to see, hundreds of tons of steel being lifted in the air like some kind of metal toy, dangling at the end of metal ropes, suspended in the sky against a backdrop of the James River and the James River Bridge. People simply stopped breathing watching it. More than one "Lordy!" could be heard.

Don Doverspike recalled when the steelworkers staged another propulsion area, the reactor space, and one of the government inspectors watched it all and told the shipbuilder, "This is cool."

Doverspike laughed. He'd looked at the inspector and then back at the assembly. "I guess you're right. I guess if you hadn't done it six or seven times." On the waterfront at Newport News Shipbuilding, lifts meant business as usual. Like the keel-laying that day—sure they were making a big deal of it, for the moment. By the next morning, though, all the VIPs would be gone, and it would be just another day on the James, another hard day's shift.

Indeed, when Mike Butler watched engine room number-two unit being dangled into Dry Dock 12, he had on his mind what they had already done on CVN 79 and what they had yet to do. Even though the day officially commemorated the laying of the keel, bottom sections of the ship covered parts of the dry dock, securely resting atop hundreds of those wooden cairn-pedestals, some enclosed in enormous temporary metal sheds the size of military barracks and tarps to keep steelworkers and steel out of the elements. Thanks to the earlier preconstruction contract, shipbuilders had already built and lifted into Dry Dock 12 a fair number of assemblies and units. In his mental crystal ball, Butler could see the erecting of the ship, lift by lift—like a time-lapse brain video. The bottom, starboard, port, aft, forward, the bow, the island—a carrier from this mass of metal. Created by the hands, minds, and spirit of America's only carrier shipbuilders. Climbing out of his bubble, Charles Holloway walked up to the top girder to stretch his legs and walk around. On a clear day like this, he could see the concrete lines of the bridge tunnels off to the east over the expanse of leafy green tops of trees. Someone had told him NASA satellites saw Big Blue from space. He knew folks saw it from Smithfield. From this height, Tidewater still retained a sense of colonial frontier about it. The blue girder that spanned the length of Big Blue appeared to disappear in the horizon. The crane swayed. Other steelworkers also got to enjoy the view, the ones who had to climb and crawl all over the crane to maintain it, its motors and equipment, often using special hatches to get equipment in and out. Everyone moved carefully, though with purpose, taking care to transition from one space to another, as sometimes it seemed you were standing there on a piece of steel no wider than a diving board hundreds of feet above a very sudden stop.

Down below, as steelworkers assembled even more pieces of the *Kennedy* in and around the dry dock, the navy executive officer of the carrier *Lincoln*, docked just down the waterfront for overhaul, happened to be cruising by the dock. He pulled over by the platen, got out of the car, and wandered over, mesmerized by the work. He called over some of the steelworkers.

"What are those big pieces of steel for?" Lincoln's XO asked.

"Those are the first steel units for the *Kennedy*," they told him.

Well, look at that, he said to himself. The first steel for a whole new carrier. CVN 79, *John F. Kennedy*. And here he was to see it. Someday, some lucky officer would command that ship.

6

SETTING PRECEDENTS

SUMMER 2016

For the next eleven months after the laying of the *Kennedy* keel, ship-builders moved at fast-forward speed, completing, outfitting, and joining ninety lifts as the dry-docked ship under construction grew layer by layer, slice by slice, from the middle out and the bottom up. Bulkheads, decks, and compartments piled on one another, a mosaic of gray, green, and rust-colored steel structures protected in some spots by a green tarp and penned in by temporary safety-wire fencing anchored by green metal pegs. Now, in July 2016, steelworkers put Big Blue to a major test with a superlift set to tip the scales at about nine hundred tons, or about the same weight as two Boeing 747s at their maximum takeoff weight when packed with passengers, cargo, and fuel. Sam Carper and the steelworkers erected and outfitted the units as planned, in a way never attempted before the *Kennedy*. Butler planned to complete the ship in 445 lifts, 51 fewer than it had taken for the *Ford* and 149 fewer than the *Bush*. Fewer—but larger, more complete, and heavier lifts—meant less work on the ship in the dock, which meant fewer man-hours to build the ship, which meant less money. Which all meant more jobs saved in Newport News. The construction zones hummed like beehives as shipbuilders raced to install pipes, pull cables and wires, and weld, grind, and rig small mountains of metal to meet one deadline after another.

On one of those sun-splashed summer days Tidewater can be blessed with in mid-August, as the rest of the country embroiled itself in the presidential campaign battle between Donald Trump and Hillary Clinton, shipbuilders focused on completing the *first* major midship superlift. The first lifts start at the middle of the ship, with the machinery decks and spaces, especially the nuclear spaces. In some ways, steelworkers constructed the carrier up and around the nuclear compartments that shield and protect the two atomic power plants. Every pipe, weld, and wingnut in that space, every tool, component, and piece of equipment, must endure painstaking reviews to make sure it meets a set of rigid requirements dating back to the Rickover days. The guarantee of those standards, at least on the materials side, rested with Warren and Sheila Outlaw.

Born in Norfolk, Warren had been working in the yard for about a decade and a half and, for the past decade, had been supervisor for the 053 nuclear materials division. Sheila grew up in Newport News, and she had been working in 053 for about half a decade. They met at work, fell in love, and got married—though both are quick to note that Warren was *not* her boss at that time.

The nuclear division tracked every bolt, wire, and nut—anything used in building the nuclear spaces—from birth to death, even after its installation on a carrier or a submarine. Steelworkers and waterfront managers ordered all the material from 053, which scrutinized each request and then each bit of material from its inventory to make sure everything aligned perfectly. Only nuclear-related material properly cleared by navy regulations could go into those spaces, and Warren Outlaw ensured compliance. His work often faced audits, solely to double-check. Even a slight mistake on the paperwork could stop the division from filling an order.

The assembly done, the nuclear space passed muster to take its place in the drydock. Big Blue showed no strain lifting the weight, and the ship section hung in midair like a half-built spaceship. On top, the middle of the lift was shaped like a small bowl, for the reactor. The bottom middle of the lift looked like an open truck bay from a factory loading dock, and pipes stuck out of the bulkheads inside like huge metal straws. The angled sides bulged with hoses, wires, and machinery. Now, for the first time, the superlift took on the shape of a ship, with the unmistakable flattened V of the keel and sides with a weird bump in the middle, as if someone had taken a house with an angled roof and turned it upside down. A series of long steel beam legs stuck out from the bottom and sides of the superlift, as if some giant were using the ship section as a pincushion. Those legs helped support and balance the section as Big Blue eased it into its rightful place in Dry Dock 12, in the forward part of the ship. The crane moved the superlift up toward the shoreside end of the dock, stopped, and then lowered it a bit, then moved it back aft just a bit, stopped, and lowered it a bit more. The crane repeated the process one more time, the whirring machinery, warning sirens, and steelworkers' whistles drowning out the rest of the waterfront din. It took all day into the evening for steelworkers to check the alignment, make the welds, and join the section to the ship.

Carper reviewed the bigger superlifts on paper with some wonderment. He trusted the planning department, but to see it happen, to see *Kennedy*'s new build plans come to life as the crane lifted these pieces of the ship—he thought, *This never gets old.* As Big Blue lifted that ship section, the crane also carried aloft all the struggles, the tribulations, the bickering, the cold, the heat, the "I don't have the material now," the "I can't get that to you in time," the hurrying, the worrying, and everything else that went into that lift—and, for one moment, it all vanished. Everything, to Carper, seemed so clear and wiped clean when the superlift descended onto the ship. Tomorrow, they would be back painting and welding and grinding again, but for now, he actually swelled with pride, but not for what he had done. He felt like a small piece in the giant puzzle. Steelworkers really made it happen. They drove the bus; he just directed the traffic and cleared the road.

As shipbuilders prepared for another *Kennedy* superlift in January 2017, some yard workers busied their minds with another past president. Pipefitter Dan Sunderland, for one, remained confused by the continued steelworker love for Obama—and the transfer of that support to Hillary Clinton—in the region and the yard. "What did they do for us?" he asked—and quickly provided his own reply: "Nothing." He saw more I'M WITH HER shirts cropping up inside the gates, so he started wearing his Donald Trump MAKE AMERICA GREAT AGAIN caps. Trump made sense to him. "He'd be good for the country," he told others. "He'd be good for defense. He'd be good for the yard." Sunderland saw nothing wrong in expressing his First Amendment rights just as those Obama supporters did. No one said anything about his caps either way. No complaints—they all just came in and did their jobs and watched one another's back. Still, everyone felt a more partisan tension in and around the yard. Somehow, it seemed more personal. By the time the presidential campaigns revved up, Donald Trump had clearly struck a similar nerve in Tidewater that he had in the rest of the country, especially among the older White shipbuilders. Judging by the TRUMP 2016 lawn signs, billboards, and posters popping up, Tidewater Trail, particularly north of Gloucester, could rightly be renamed Trump Trail. Old-time shipbuilders circulated a Facebook post:

The IRS has returned my tax return to me this year, and I apparently answered one of the questions incorrectly. In response to the question, "Do you have anyone dependent on you?" I wrote, "9.5 illegal immigrants, 1.1 million crack heads, 3.4 million unemployable scroungers, 80,000 criminals in 85 prisons, plus 850 idiots in Washington." The IRS stated the answer I have was unacceptable. I then wrote back, "Who did I leave out???"

Yard Trumpicans rejoiced when Trump won the presidency, and the new commander-in-chief hammered home some of those themes when he appeared in Newport News on March 2, 2017, aboard the *Ford*, and spoke to sailors and steelworkers. Donning a leather flight jacket often reserved for naval officers and sporting a tan baseball cap with the carrier *Ford* insignia, Trump bellowed, "We have a great MAKE AMERICA GREAT AGAIN hat, but I said, 'This is a special day. We're wearing this.'" Looking out over the ship, he said, "This is American craftsmanship at its biggest, at its best, at its finest. American workers are the greatest anywhere in the world." They applauded. "We'll have more coming." More applause.

Then he said of the carrier, "The same boat for less money. The same ship for less money." Folks had to overlook that one. The *Ford*, topping $13 billion—the most expensive carrier ever built—was still running behind the updated schedule and over budget. "I just spoke with navy and industry leaders and have discussed my plans to undertake a major expansion of our entire navy fleet, including having the twelve-carrier navy we need." No way the audience there would not roundly applaud this, but no one there knew at the time that Trump's Pentagon was already planning for a future with fewer flattops.

Just a few months after Trump spoke on the *Ford*, as the country grew accustomed to its new president looking for ways to divide Americans, Newport News Shipbuilding itself boasted of a new president, starting in the summer of 2017, who made sure no such divisiveness made its way to the Tidewater waterfront. The contrast between the new yard chief and the commander in chief could not have been starker.

When you grew up as a sports-minded kid in the 1970s in St. Louis, you certainly wanted to play Little League baseball. With players like Bob Gibson, Lou Brock, and other Hall of Famers–to be, it became only natural for little boys. Little girls could not play Little League ball—until Jennifer Roman came along. She wanted to play Little League, and little else mattered. As a tomboy and the middle kid in a squad of five, she learned early on that you took want you wanted, or you went without. Diminutive but tough, she held her own. Being kid number three, she also learned how to take a lot of "constructive criticism."

She definitely had her share of "criticism"—teasing—as the only girl playing ball in Little League. But the sense of greater integration all around in St. Louis made it easier for Jennifer. Her beloved Cardinals had been pioneers in fielding Black players, but Jennifer's community took that spirit even further, integrating her school and even the jazz band she played in. She attended what would be considered African American schools, and she was biracial; her father was first-generation American, his family coming from Puerto Rico. He had moved his clan out to St. Louis to take an engineer's job at McDonnell Douglas.

Born with an engineer's genes, she of course knew the stats of her favorite Cardinals. Something else caught her attention and got her thinking. She saw a picture of Cards coach Whitey Herzog atop the sports page, scratching his head and rubbing his chin on the first day of spring training during the players' strike. Until that moment, she had never really thought of the manager's role. Realizing that he'd built a team with all his players on strike, she asked herself, *How do you manage that?*

As her St. Louis childhood ended, Jennifer Roman became more concerned about the management of her own life. Excelling in math and science, she persuaded herself that she wanted to be an engineer. For whom, doing what, she didn't know, but she first focused on picking a place to study. Being from St. Louis, she first thought of the state university. The US Air Force and US Merchant Marine academies both accepted her. Her parents pushed her toward the latter, which offered not only classroom education, but also real-world experience on ships at sea. She agreed to try it and fell in love with just about everything connected with the school. Hers was not the first class of women plebes at the academy in 1982; it was the ninth. She learned what made ships tick and traveled the world making them work. But she also learned something that served her well as she rose through the ranks of the maritime community: a small number of people, connected, committed, and dedicated to one another, supporting one another, could be extremely successful.

At the academy, Jennifer Roman met and started dating her future husband, Blake Boykin, who then served as a soldier at Fort Eustis, just north of the Denbigh section of Newport News. She applied to Newport News Shipbuilding, partly to be close to him, but nothing came of that first interview. She settled for a job at the David Taylor Naval Research and Development Center, later called the Carderock Division of the Naval Surface Warfare Center in northern Virginia, where she tracked data for a guy who held patents on propeller designs. She and Blake spent a year doing the long-distance-dating thing. She reapplied to the yard and this time got the job, in 1987. She took the last name Boykin when she and Blake married about two years later.

She started as a nuclear engineer at a shipyard satellite building in Oyster Point, the growing commercial "new center city" area on Jefferson Avenue, so she avoided the kind of waterfront attitudes other women had to face behind the gates. One policy at the time needed to be changed: it prevented women from going into potentially high-radiation areas, due to concerns over the risk to pregnancy. In her department, folks worried most about making sure everyone carried their own weight and helped the team.

Jennifer Boykin found the same sentiment when she entered the shipyard proper the following decade. The seasoned shipbuilders gave the woman engineer a chance to prove herself, to see how she contributed. The former St. Louis Little Leaguer knocked it right out of the park—as a team player. Shipbuilders never cottoned much to solo superstars.

Boykin entered the carrier shipbuilding program on the Newport News waterfront as construction superintendent on the *John C. Stennis* in the spring of 1994—the same spring, as it turned out, that the US Navy first put women on carriers when more than five dozen served aboard the *Eisenhower*. As the yard finished the *Stennis*, one of the pipefitter bosses gave her a small piece of copper pipe, about three inches in diameter and a quarter inch thick, that he had cut and polished, telling her, "This is to remind you where you started." To help keep herself grounded, she kept it on her key chain as she ascended to the top of the yard's pecking order.

After *Stennis*, she moved on to be a superintendent for *Truman*—a special job to a native Missourian. Mike Petters, the *Truman* program manager, took note of the upcoming supervisor. She saw him as a mentor. For both carriers, she managed the traditional outfitting work for most of the ships' spaces other than for machinery—the ones for berthing, offices, the hangar bay, and the galley compartments. She took care of all the spaces where sailors would live and work, and in doing so, she developed some strong and mutually respectful relationships with navy department heads who went on to become admirals.

Boykin, meant for bigger things, moved up to run the yard's facility division, where she oversaw, among other things, Big Blue's lift-power surge to 1,050 tons. She also played the role of community ambassador, and that left a more telling mark on her psyche. While working as superintendent on the *Stennis* and the *Truman*, moored in the south end of the yard, she noticed how trucks delivering some of the material for those ships blocked the very narrow roadway to and from the pier area. "If there's any kind of injury or emergency, an ambulance could never get through," she warned.

The yard needed to expand the road and the property rights there, which meant buying up old, mostly abandoned homes. In one, though, lived a gentleman in his eighties, who did not want to be forced out by the "big, bad" shipbuilding company. Instead of leaving all that to the lawyers, Jennifer met with the homeowner and explained the situation. When she had finished, he took her hand, shook it, and said, "I'm so glad you're the face of this company. And I'm ready to go live with my nephew now."

The face of the company became increasingly important to the Boykins as they settled more comfortably into Hampton Roads. Jennifer Roman had remained lukewarm about the region after she first moved there. She'd grown up in St. Louis, with its big-league sports teams and big-city culture. "Newport News didn't offer any of that," she lamented. "It never dawned on me to live in a region that does not have one professional sports team. In St. Louis, you had baseball, hockey, basketball, and sports fanatics. Newport News, Norfolk—the whole region felt kind of small. You didn't have anything like Broadway or operas or musicals."

She wasn't the first to notice this lack. At the end of World War I, a couple of soldiers, Corporal Willie Shifrin and Sergeant Hal Oliver, wrote a song, "Newport News Blues":

Newport News blues will surely drive you mad.
You start into jazz, then you raz-ma-taz.
Oh, way down south in the land of cotton,
Your Uncle Sam has not forgotten.
You're a-way, a-way, far a-way from Broadway.

Yard management in the past had tried to change that, but she wanted to try harder.

Her promotions made that possible. When Petters became yard president, while Northrop Grumman retained control, he promoted her to executive vice president, one of two for the company at the time. She took over management of yard continuous improvement, which included quality assurance. Yard president Matt Mulherin put her in charge of a new overall engineering group in 2012. When Mulherin retired five years later and Boykin became president, the outgoing yard chief gave his successor one of the early shipyard badges—a pin, really, with a small photo of the worker. Dating back to the 1940s, this pin featured a picture of a woman. Some of the shipbuilders she knew treasured such yard pins, worn by their great-great-grandfathers, but she had never seen a woman's pin before, and she relished the gift.

Petters remained extremely pleased with the person he'd entrusted with the management of the shipyard. "Sure, she takes my advice," he explained. "Then just ignores it and does what she wants, like she did back on the *Truman*." She nodded at his suggestions and smiled and acted gracious, and then she did what she, in the end, determined to be best. She retained one gift, though, that made her an outstanding shipyard president: the ability to look past the horizon, to consider everything about the yard—its culture, its technology, and its programs—beyond the immediate, beyond the tactical and toward the strategic.

But Boykin also learned a great deal from the people who reported to her. Whenever she took over a new division or a new part of the business, she came quickly to learn from and rely on people who already understood that part of the business. While senior leaders showed her the art of leadership, the shipbuilders taught her the art and craft of shipbuilding. Those shipbuilders, mostly male old-school traditionalists, all came to hold Boykin in their highest esteem. They didn't see her as a woman boss. They saw her as an effective boss, a smart boss, a good boss. And Boykin showed them the same respect. She grew up professionally on that waterfront, never shying away from the hard job. She sought tough leadership jobs no one else wanted and volunteered for the difficult tasks. When she found them, she leaned in. She searched for the same in yard up-and-comers. Petters saw and appreciated the way she collaborated, the way she built strong teams that seemed to gel—and delivered exceptional results. At the ripe old age of fifty-three, she became not only yard president, but also corporate VP for Huntington Ingalls Industries, directly reporting to Petters.

The first female president of the yard remained more interested in breaking new ground than smashing glass ceilings. The way she saw it, manufacturing all around the world had started to undergo a new kind of industrial revolution, and she saw digital shipbuilding as the key to success for her business.

With a tablet tap, shipbuilders pulled up work instructions and training videos. Eventually, workers anticipated getting a little pop-up window through which they could contact the designer and engineer, showing them an image to get the best technical help, right through the tablet, which also tracked the amount of time needed to complete the job. It would be like a house general contractor being able to co-ordinate the work simultaneously with the Sheetrock guy, the electrician, and the plumber, all on the tablet.

Boykin and the yard invested in more than just the digital ship-building revolution, spending about a billion dollars during the previous half decade on its facilities, with plans to put out another half billion in the coming five years. Some of that money directly bene-fited the workers themselves—like the efforts to provide more park-ing, upgrade restrooms, and create more eating and meeting places throughout the yard.

The former "math and science kid," who had been the first girl to play Little League baseball in St. Louis, did not forget her roots, either, as she climbed up that corporate ladder. She helped start a local program for young girls called Girls with Engineering Minds in Shipbuilding, or GEMS, which would boost middle-school girls' confidence to stick it out through math and science classes, with the help of women ship-builder mentors. The girls journaled about GEMS experiments. Each also wrote two letters, one at the beginning of the year and another at the end. The change between first and last letters particularly struck Boykin. Often they would write something like "I'm not good at math, I'm not good at science, and I don't think I want to do this." At the end of the year, they would write, "This is for me. I'm sticking to this."

Boykin expanded that effort, working with the local educational mainstay, Old Dominion University, to develop Project Vision, partnering with the city to build a new office building at the poor East End of Newport News. She planned to house yard employees—digital designers, engineers, planners, some IT people—on the second, third, and fourth floors. The city, though, owned the first floor, operated by the university for both workforce development and STEM (science, technology, engineering, and mathematics) outreach. The East End students had their own place to learn some basic programming and computer-aided design (CAD) skills to level the playing field and increase the number of women and minorities in the STEM fields.

She saw STEM as the great equalizer, not only for local kids, but also for the region. She sought to make Hampton Roads a tech powerhouse, attracting all the best working and teaching talent to its colleges, companies, and the rest of the community, to unify Tidewater around that idea so the region could become a modernizing force in the state and throughout the country.

However, Boykin put such far-reaching plans on the back burner while she focused on her more immediate concern—CVN 79. On her first day as president, she named seasoned shipbuilder Lucas Hicks as the vice president for the *Kennedy* program. He promptly took on CVN 78 work as well, to help fix all the problems with the *Ford*, overseeing both critical shipbuilding programs.

Hicks had worked the Newport News waterfront for more than two decades, most recently as the *Ford* superintendent. Shipbuilding was a Hicks family business. His father spent his entire working life at the yard. His grandfather started there as an apprentice. His great-grandfather started there when he was only sixteen, eventually becoming a riveter foreman.

Instead of heading straight into the yard out of high school like many other career shipbuilders, or going to the Apprentice School, Lucas Hicks earned a degree at the Merchant Marine Academy in

New York and started at the yard afterward as an engineer working on the navy's Los Angeles–class attack submarines, also called 688-class subs, after the hull number of the first one, the *Los Angeles*. He'd spent decades honing his shipbuilder skills and instincts, working on the design-and-build team for the *Ford* just as the yard started to use its new digital production tools.

After taking on the *Kennedy* and *Ford* jobs, he gathered up the master shipbuilding experts, who knew the ships best, and asked them to "tell me how you're going to get the job done and what you need from me to get us there."

Concern continued to grow about the *Kennedy*. Despite all the benefits Butler, Doverspike, Carper, and the rest of the shipbuilders gained from digital engineering, the yard still struggled to meet its promised man-hour cuts. In July 2018, the Pentagon's naval warfare division sent a report to Ellen Lord, the Defense Department's chief weapons buyer, saying the shipyard was unlikely to "fully recover the needed eighteen percent" reduction in man-hours on the *Kennedy*, claiming that Newport News would be unable to fulfill its top-priority promise and chief contractual obligation for CVN 79. That did not bode well for the future of the *Kennedy*, the carrier program, or the thousands of steelworkers on the waterfront.

7

TRADE WARS

FALL 2018

Big Blue, Charlie Holloway, and scores of riggers prepared for another superlift in late September 2018 to mate the bulbous bow—the minisub-like bottom meant to stop the carrier from nosediving as it broke through waves as tall as houses—to the rest of the ship. As during other lifts like this, workers lined the side of Dry Dock 12 under a cloudy fall sky to watch the monstrous crane heft the rust-colored 726-ton lower bow over the ship quarry and move the section ever so slowly toward the midsection of the CVN 79 hull, now starting to take on the true carrier shape, with its beams, bulkheads, and decks, manned now by welders, machinists, and other shipbuilders waiting to put this part of the ship's underwater hull in place, adding another 122 feet to the *Kennedy* keel. One of those watching was *Kennedy* machinist Aaron McCann, who'd started to work in the yard that year. He watched the new carrier bow hover overhead. "Awe-inspiring," he'd tell others later, "like watching a large house jumping up in the air and flying."

With the bow completed, viewed head-on from the yard waterfront, the ship looked like a giant metal crouching crocodile, secured by a line of giant poles, jacks, shims, and girders. Big Blue needed to flex nearly all its newfound muscle to fit the steel croc with a new snout—a massive nose job sixty-four feet high by fifty-four feet wide, filled with eighty-four thousand square feet of tanks and lockers, all packed with more outfitting than any such section for previous ships. Don Carper watched and smiled with the achievement of another milestone, the completion of CVN 79's underwater hull and its structures from the main deck down.

Now painters, welders, and other trades workers boarded the ship, climbed through its compartments and began some of the most physically demanding jobs—back-breaking, spine-twisting, muscle-numbing tasks. They needed to work harder than ever. The yard still had that contractual obligation to get it done with nearly a fifth fewer man-hours than on *Ford*.

And so, coming from different compass directions on Route 17, Mike Butler began his early morning commute from near the Virginia–North Carolina border, and the Murphys and Elliotts started off from Gloucester and Mathews in the Middle Peninsula.

Now, it would be natural to assume that those who lived closer to the yard, in Hampton or in Newport News itself, would be able to sleep longer, thanks to the shorter commute—a wrong assumption. Newbie machinist McCann, for example, lived just a few miles away from the yard as the crow flies, in Hampton, along with thousands of other steel-workers. Thousands of others lived in Newport News—folks like Jordan Patterson, who'd joined the yard, as a painter, about the same time McCann did, a few months before the bulbous-bow superlift. They lived just a few minutes away, yet both young men woke up and started their morning commutes about the same time as those who lived much farther away. Unlike Charlie Holloway, who could walk to work, or others like Butler, Doverspike, and Carper, whose gravitas earned them reserved spaces, the McCanns and Pattersons faced a daily battle for a good space, one of the most stressful moments of their days, at the Hidens Complex, where most of the rank and file parked.

McCann and Patterson steered their cars along the dark Newport

News streets, crisscrossing along the front of the yard to get going the right way on one-way Warwick Boulevard, out near the old *Daily Press* newspaper building, to gain the right entranceway to the Hidens Complex, a massive dirt, rock, and gravel parking lot near the northern edge of Newport News Shipbuilding and a few blocks from the waterfront. Workers found the traffic almost bearable in the morning, with *relatively* little battling to get to the yard or even to Hidens. The real fight started over the choice spots in Hidens where one could make the speediest exit out of the lot at the end of the day. To even have a shot at nabbing those first few spots, steelworkers like McCann and Patterson had to arrive more than an hour before their shift started. Most failed, and they banged their steering wheels and cussed, forced to settle for spots farther away from the exit.

Once safely anchored in Hidens, McCann and Patterson joined the streams of steelworkers for the fifteen- to twenty-minute walk to the gate, lugging skateboards, pushing bikes, or carrying running sneakers. In the fall, some found the walk pleasurable, but come winter, that freezing wind off the James cut like an icy dagger. Many workers remained in their cars to sleep for another hour or so instead of going through the turnstiles so early. Like a parking lot the night before a big concert, carloads of people stretched out in the backseat, reclined in the front, or set up camp in the back of a pickup. Near the gates, steelworkers lined up at the food trucks for breakfast. If the Bojangles truck line didn't stretch too long down the street, steelworkers might stop for a bite, although some preferred Eddie's, a stalwart one-story greasy spoon that had stood outside the gates for decades.

Patterson usually found breakfast lines ridiculously long. Like many others, he usually marched through the front gate right away, a little after six. Workers clustered and cackled about the day's work, last night's Nats' game, or family life. About six forty-five, they'd head over to get their shift checks, which marked them as officially being in the yard for a shift's work, and waited for the first bell to ring about fifteen minutes later, when they met with their respective foremen, stretched their limbs, and got their check cards detailing the work the yard expected them to complete that day.

The yard required foremen to lead their workers through their stretches before the morning shift started, usually as the sun's rays first started to filter into the waterfront. The tool room opened at 7:30, and the foremen's instruction to trades workers usually lasted only a few minutes, theoretically leaving plenty of time for the mandatory limbering-up. Compliance varied. When both foremen and their workers opted not to comply, no one had a problem. When workers wanted to stretch and foremen didn't care, that presented no problem either. But when a foreman demanded compliance and the workers refused, the grievances started flying.

The yard required foremen to write up workers who failed to comply. Wayne Deberry sounded alarms in November 2017 when the Local 8888 leadership agreed to help the company start fitness programs. "I knew this would happen," he later told workers. But then the International Steelworkers and HII inserted language in the contract to *require* employees to stretch and exercise before starting work. No spate of muscle cramps or similar ailments prompted the stretching mandates; the requirements arose as part of a national effort in industrial sites for such exercising. The wording in the contract said *voluntary*, but it worked out differently. Deberry and other union officials almost immediately filed grievances as steelworkers pushed back. Many veterans failed to see what in the hell this had to do with shipbuilding. The docking continued, and the number of grievances increased.

So thousands of workers lined up and touched their toes, reached for the sky, and *stretched*. The yard expected everyone to participate— Big Ed, Wingnut, the Murphys—everybody, even someone like Erica Brinson, an X-36er who worked in the tool room. She worried more about not having the tools the steelworkers needed and being cussed out when she failed to provide them. As a union representative, she returned fire every time. She would be one of the first to say, "They can't make you do that," if some company requirement violated the contract. The stretching did not. To comply, some workers did the bare minimum. Younger workers minded less; many of them only wanted to keep their jobs. They still remembered the recession. Besides, many found the stretching easy enough.

Painter Jordan Patterson gladly touched his toes and twisted his body this way and that. These stretches saved his thin Black body from more than one cramp. He folded, pressed, and otherwise mutilated his body for six hours at a time, and he appreciated anything that kept him from hurting himself. Let other folks complain. He was content to stretch and keep his job.

Most young people in Hampton Roads felt the same way. "You pretty much know you're going to wind up at one of the big three around here," McCann explained one day at the union hall, telling other steelworkers what they already knew. "There's the military, Walmart, or the yard. That's how I wound up here." He entered the gates in January 2018, in skeletal reconstruction, cutting steel for a living.

He focused on his morning safety brief, his job assignment, and job sheets that spelled out his work that day, to make sure the task and the instructions matched exactly. Then he went to his metal locker, like an ultra-big footlocker, to grab his tools—multiple grinders, punches, and hammers, lots of heavy, bulky equipment, and of course his work tablet. He selected tools he needed for that day. While many steelworkers packed their tools into bags, McCann preferred to use a bucket, which he found easier to carry through hatches and up and down ladders. Getting ready to descend from one level to the next, he attached a rope to a strap he had fashioned for the bucket handle, carefully lowered it down, trying to minimize the back-and-forth circling and swinging until it touched the floor below. Then he climbed down the ladder. Hopefully, no painters planned to invade his space today, as they both could not work in the same immediate proximity at the same time—either the metal-cutter or the painter had to call it a day.

Jordan Patterson started in the yard, as a painter, about a month after McCann started his job. "I had nothing against painting," he told other steelworkers, "but I just wanted to do something else here." But following the birth of his daughter, after nine years of trying to get hired as a welder, he wanted any position inside the gates. Quite a few from his family already worked there—aunts, uncles, and cousins—and they all told him, "You need to apply again."

He survived a couple of interviews, but he figured he messed up. The yard interviewer asked him, "How high can you go?" He acknowledged a slight fear of heights. On a carrier, shipbuilders knew they could count on working up high. "I could stand up on a skyscraper and I'm not afraid at all," he said. But walking across a rope bridge or something like that, it didn't matter if he had to look down only eight feet—he froze. He later learned to conquer that fear, and the yard needed to hire more workers in a hurry, so he became a painter. "When I have to work up high, I just deal with it."

His work instructions in hand, he retrieved the necessary tools from his locker. His locker held grinders—seven-inchers and three-inchers—as well as a handful of individual brushes he used to make the job look pretty. Today, he also might need his self-modified three-inch roller to reach those weirdly angled spaces. Or he might need his seven-inch roller, its handle bent to look like a huge number 7. He chose from a dozen different brushes, all bent this way and that, to be able to paint whatever they needed him to, in any tight spot. Whether he needed to descend into one of the tanks or climb way up in the island, if they wanted it painted, he got to it, sometimes with a brush, other times with a sprayer.

Often the most difficult part of his job was trying to determine exactly what they needed painted. One time, he got a work order to paint a microwave in a certain space. But when he got there, he found seven microwaves and no further instruction on which one to paint. They often gave him wrong ship coordinates. Builders broke down the carrier layout into digital coordinates, based on deck, section, and other determiners, like a map grid—the only way to track all the trades work being done on a carrier at any given time. Patterson often arrived at a particular coordinate on his work order with nothing in the space matching the order. If he painted the wrong thing, that led to all kinds of other trouble.

Most days, like that day, he arrived at his space, checked his instructions, grabbed the right brush—today, a three-incher—and crawled into a cabin just a bit bigger than a coffin, his workspace for the next four or five hours. In the yard, claustrophobia threatened more careers than acrophobia. Workers sometimes lied to the company or even themselves about their ability to stay stuck in small, tight places for a long time. Freeze for a moment in that situation, and you could kiss your yard job good-bye. Workers likened getting into some of these tanks to spending a day inside a tiny trash can. To determine whether workers were claustrophobic, the yard made them shimmy through a tight wooden tunnel, carefully monitoring their progress and reaction.

Shortly after Patterson squeezed into his coffin-cabin, he appreciated his morning stretching. He crawled out, limbered up for another minute, and climbed back in. Some folks used that short break to rest for a bit, putting a bucket up against a bulkhead and leaning back—bucket-sitters, in yard parlance. Patterson found that position uncomfortable because it lacked any real lumbar support, but he noticed other workers falling asleep sitting like that with tools in their hands.

Meanwhile, McCann worked on an incremental fitting, connecting machinery to a bulkhead, first ensuring the proper preparation of all the steel. Other steelworkers had joined, assembled, or stored the steel pieces and components outside, so McCann and other machinists first often needed to blast the metal clean of rust and dirt, then add a little paint protection before moving the pieces into position. Metalworkers then ground the edges so as to better join the pieces and sometimes tack-welded them into place. Depending on the job, hotworkers then lit their torches to heat the steel, sometimes two inches thick, often heating the whole metal section and sometimes carbonizing the ashy area they had just ground. They did this, space after space, by the thousands—painting, cutting, and manipulating metal, over and over again. Over and over was the essence of building a carrier. Sweaty, parched, and miserable, you smelled yourself most of the day.

McCann checked the space to see how much room he had to maneuver, and he scrutinized the metal connector piece. He made some calculations and cut the piece with his black diamond drill, sparks flying. Now he cut through half-inch plates of steel stuck together at all kinds of different angles, and he couldn't take the whole shift to get it done. When he finished, though, he had a shiny piece of metal, ready to be painted over. His jack sheet told him, "Direct painter," but the supervisor had failed to give the right coordinates, and McCann waited there hopefully for the painter to arrive.

McCann decided to contact his foreman by cell phone—often the best way to reach a waterfront supervisor in the yard. Best, but not reliable. The shipbuilder stood there in a metal box within a maze of big metal boxes within a maze of bigger metal boxes in one huge metal box, and he might or might not get a signal. It was a coin flip. It was a double coin flip, because while he might be able to get a signal, his foreman might be in a place without one. And forget about making an actual call; the noise made it impossible to talk or even listen to a voice message. Many shipbuilders in the yard used private cell phones to text one another.

Of course, the yard prohibited waterfront workers from using personal phones during work hours. The yard also prohibited workers from listening to iPods or other devices when making their way through the yard. Supervisors tended to look the other way when folks used their phones in a break area, though. The yard declared earbuds verboten, which rankled many. However, many jobs required earplugs, and determining whether a worker was wearing earplugs or earbuds, especially when someone, like Patterson, was outfitted in a hood or other headgear to safely do the job, was difficult if not impossible for most supervisors. Many steelworkers listened to their devices on the sly, a soothing thing when dealing with some of the daily work annoyances inside the gates.

Patterson, for example, found that some trades workers cared little about the condition of a space they left for painters. He had entered many a space, brushes in hand, and discovered a chaotic mess, the metal in some cases damaged by a welder's carelessness, and he spent half his time just cleaning the steel so he could start the actual painting. Sometimes, another trade would come in, install a new box, and ruin a perfectly good paint job, forcing him to repaint everything he had just finished a day or so before. He also often found some resin patches throughout the space, requiring him to ball himself up for two to three hours scraping that mess off; the grinder proved absolutely useless. Uncramping his hand, he'd silently curse the steelworker who left the space like that. Patterson often retouched the same space four times due to sloppy work by other trades. He'd shake his head. *You'd think after so many carriers they'd be able to plan this better.* It all seemed so backward. "It's like they want you to paint something just for someone to later mess it up and for you to go back and repaint it," he complained later at the union hall. Most of the paint protected metal that had just been ground, welded, or cut, in which case, he told folks at the union hall, it made more sense to spread a temporary gel over the bare metal for protection.

But for now, he just repainted. First, he prepped the metal, cleaning off every last bit of soot and debris. He checked every weld to make sure the hotworker had tacked it just right; he needed to avoid painting over a bad weld. If the weld looked somehow wrong, he needed to contact his foreman, which, as noted, often was difficult. His foreman then needed to fix the situation with the welder's foreman to get the weld done correctly, requiring more iffy communications. The way daily tasks lined up at the yard, Patterson could not move on to some other job in another space, because it would be filled with other trades workers doing their daily tasks. Worse, depending on how long it took to resolve the wonky weld, the prepped metal could easily start rusting; it happened very fast in that salty coastal air. It would need more prep later.

While worrying about their own space work for their respective trades, both painters and metalworkers also kept their eyes peeled for another force that could throw a wrench in all their efforts: the men and women from X-11, known throughout the yard as the "foundation people." A kind of quality control SWAT team, X-11ers just popped up, as painters, welders, or metalcutters got about halfway through a job, and say, "No, no, no. Stop! This counter has to go from here to over there." Then the trades workers would watch their spaces being pulled apart and all the components being moved around or even moved out.

That really aggravated Patterson during a job that required him to wear his "bunny suit," the white total-body hooded vinyl outfit that held heat like an industrial oven. Wearing that through a shift cooked him like being baked alive. He sometimes needed help from other painters to peel his clothes off him like a banana skin.

McCann struggled enough getting to the job space. Shortly after he started working in the yard, he was climbing up a ladder to cut some metal for the *Kennedy*. He secured his bucket. The ladder held steady and was well anchored. He did everything exactly as required to get off the ladder safely, following official yard procedure. But when he stretched his body, he overextended, and his work boot locked into the ladder rung, causing way too much torque on his knee, twisting it about ninety degrees, dislocating the kneecap, and tearing the insides apart. After that, he was at the mercy of the "team from Walmart"—

the outsourced workers compensation legal representation they had to use.

However, the chief cause of daily friction among painters, metal-workers, and the other trades remained the rules for how they could work—or more accurately, could not work—with one another. Paint-ers could work within ten feet of hotwork (jobs involving blowtorches, electric arcs, and other tools that produce high heat or sparks). But paradoxically, hotworkers had to remain at least thirty-five feet away from paint work and combustibles, to create a safety zone so flying sparks couldn't cause a fire with the highly flammable painting ma-terials. That rule applied not only in the same space, but in adjacent spaces, too. If a shipbuilder was painting a bulkhead, for example, no one could be doing hotwork on the other side. Metal, after all, con-ducts an electric current like that from a welder's arc, and sparks can pass through very small openings. Now, under certain conditions, hotworkers did cozy up to painters, but the yard had ingrained it in the skulls of painters and hotworkers to err way on the side of cau-tion, so they generally kept thirty-five feet away from each other. Don Doverspike and other managers made sure their people remained deadly serious about following that rule. It created extra work for managers like Doverspike, who sequenced the work in the thousands of spaces to separate the trades. But try as they might, Doverspike and others failed to keep the two trades from crossing paths. More than once, Big Ed Elliott arrived at a space, saw a painter, and found a bucket to sit on. "Might as well get comfortable," he'd say. "Not going to get any work done here today." Metalworkers understood the rea-soning. When they were bent over in a tight space with their hoods and safety glasses on, earplugs in, grinder growling, sparks showering all over, they might not even notice a spark. They operated like deaf horses with blinders on. Teammates kept an eye out, of course, but they didn't see everything, and fires erupted with match-flick speed.

It could be so damn annoying when a painter like Patterson would arrive in a space where steel was being cut, ground, or welded—and open a paint can twelve or fifteen feet away. That served as a fine distance for the painter, but it ended a shift's work for a hotworker the moment the paint lid came off.

The hotworker would ask, "Are you kicking us off our job?"

The painter would reply, "I'm cool within ten feet." He or she stayed, but the hotworkers had to stop. All workers had their assigned tasks, and they needed to account to their own foremen for their work that day. If a painter could do the job ten feet or farther away from the hotwork, then that painting had to be done. The fact that the painting forced the hotworkers to stop their job—well, that was their problem. Efforts to change the rule to thirty-five feet for everyone failed, and during the *Kennedy* dry-dock work, the two separate distances applied. Some hotworkers threatened to put a sheet of rubber over the paint buckets, but they knew the yard forbade such an action. Still, the rules and practices delayed work and added man-hours, creating more problems on the *Kennedy*.

After twisting their bodies, breaking their backs, and engaging in trades wars all morning, workers readied for lunch. A generation ago, yard workers enjoyed an hour for lunch, and with all the fast-food places, bars, and go-go clubs right outside the gates, that was more than enough time to get their fill of what they wanted or thought they needed. The yard cut the lunch break in half but ended the whole shift a half hour earlier, too, saying the steelworkers would rather get home earlier than spend more time on lunch. But no waterfront workers remembered it that way. They believed the change had more to do with the yard's desire to keep its workers away from booze and sex. In the bad old days, when drugs and crime ran rampant nationwide, shootings took place right outside the yard, bullets whizzing right through the gates. Many steelworkers thought the lunchtime cut came as a directive from the US Navy, which wanted to clear the waterfront of midday distractions to keep sailors as well as workers in line. A half hour left no time for any extracurricular activities, and the bars, dance clubs, and sit-down restaurants moved or shut down. A couple of Asian and superfast takeout joints remained, converting to assembly-line service for the few workers who ventured outside to shovel down food and guzzle a soda.

Lunchtime on the waterfront was different a century ago. Vendors sold food from horse-drawn wagons inside the yard. One wagon traveling along the waterfront was used to transport foremen and superintendents, the "brain wagon," which carried the yard royalty around like kings. Now, the only kings in the yard were the Kings food trucks allowed to park at a few locations. Quick kebobs and so on. Some workers, when they had the extra few minutes to head outside the gates, went to 7-Eleven or Eddie's.

Even fewer workers on the half-built *Kennedy* in Dry Dock 12 found the trek outside worth the effort. Painters, welders, and all the other trades workers had to be in their spaces until 11:59:59 to turn over their shift tickets right as the digits hit 12:00:00. Those tickets identified them as being on the clock at their stations until they handed them to their supervisors. Once steelworkers turned in those shift tickets, they began to move, going from turtle speed to Flash-like zip. Depending on where they were on the *Kennedy*—what level, what space, and in what state of metal mayhem—it could take ten minutes to get off the ship, out of the dry dock, and onto the pier. Leaving another ten minutes to get back onto the ship and into their workspaces, those workers had only ten minutes to get some sustenance, for they *had* to be back on the job by 12:30:00.

Some risked it. They fast-walked through the yard and out of the gates to one of the few places they *might* be able to get something they wanted. Full-grown men and women, some the size of Olympic heavyweight weightlifters, streamed through the yard with hard hats, safety vests, and work boots on feet, moving like little windup toys. They wanted to sprint, but the yard forbade running inside the gates.

The yard also banned vehicle traffic inside the gates during the lunch half hour, and even those supervisors who had bikes couldn't ride them. All for safety's sake.

The truly intrepid hit the food trucks outside the gate or the few parked just inside. You stood in line there, hoping that (a) you got to order and get your food in time, (b) the trucks still had something you wanted to eat by the time you got to order, and (c) it was edible.

There were no benches or picnic tables along the waterfront inside the gates, where they would have provided postcard views of the James. The yard had no real cafeterias. The CVN Carrier Café at the Innovation Center, on another part of the waterfront, was the only place with any kind of tables. *Kennedy* steelworkers who raced over there would often find the seats already taken by the designers, engineers, and others who worked in the center.

By the time they hoofed it over to the center, though, they'd committed themselves to get something at the canteen if they wanted to eat at all. By then, workers from the Innovation Center or one of the many contractor team members who daily infiltrated the yard had already snatched up the good food. They enjoyed more flexibility as to when they started lunch, and most began their break just before the steelworkers' noontime rush.

Given all this, most steelworkers grabbed something quick from the vending machines or like, Patterson, brought their own food. Patterson sometimes skipped the meal by filling himself up with the water always available on the ship. Same with ice in the summer. A few lucky steelworkers, like Erica Brinson, worked in places like the tool room and didn't have to worry about running out and getting a meal. They had a microwave and a refrigerator in the room. Sometimes, they even ordered out.

It would be interesting to see a time-lapse video showing the impacts of the yard's half-hour lunch on the nearby Newport News downtown business front, moving from a vibrant worker-packed street scene along Washington and West streets to the sparse near-ghost-town silence that *Kennedy* workers witnessed. Restaurants emptied and disappeared, leaving behind law offices, boot shops, and new Newport News Shipbuilding satellite offices. Crumbling empty lots multiplied, rented out for expensive parking. Some sat in their houses and stared out the window, just waiting for some foolish steelworker to illegally park on a little spot of grass next to their houses so they could call for a tow truck.

On slow afternoons in the union hall, steelworkers debated whether the yard should move back to an hour's lunch. Most thought it a good idea. Some might drive somewhere for lunch. Of course, that might revive yard concerns about workers drinking during lunch, but the workers noted that no one needed a longer lunch break to sneak a shot or three. If people wanted to bring drink or drugs into the yard, they'd find a way. Most on the waterfront thought those worries about drugs and alcohol were overblown. The steelworkers knew they could be tested anytime. If there was any kind of accident, for example, *everyone* who was there at that moment would likely be tested—those involved, those who witnessed it, and those who just happened to be walking by. Hell, if you tripped walking along Shipyard Drive, the yard had you tested, with zero tolerance for the wrong results. They scraped hair from workers' legs for tests, and workers believed the yard could determine traces of some substances for up to seven months.

Also, the longer lunch might, just might, improve the parking situation. Some workers might move their cars around to make it easier to get out when the shift ended. That alone would make the change worth it, to avoid the chaos and carnage when the shift-ending siren told most of Newport News it was three-thirty. Then thousands of steelworkers rushed toward the closest exit from their space to get off the ship and out of the dry dock, like the last day of high school before a summer break but bigger and happening every workday.

They sprang like racehorses at the starting gun, some running, some skateboarding, and others cycling—all in a mad rush to get to their cars and be among the first to squeeze into the bottleneck to leave Hidens. McCann watched the hordes and thought at least ten thousand workers must be heading down to the lot at that moment. Some van-pooled—vans can get great parking spots, a gold ticket. But you never knew when the driver could be sick. Some shelled out the ten bucks for paid parking spots, but most refused, on principle.

Runners, skateboarders, and cyclists raced down Warwick at the same time bouncing over railroad tracks and no bike lanes to protect them. Dozens of buses ran from various spots in the yard to Hidens; they merged with cars from other parking spaces along the waterfront, all converging on one road, all merging to the right to get into the one lane that would enter the parking lot. Workers timed it to the minute—when they needed to be in the car, key in ignition, and wheels kicking up dirt. They had no time to waste on the air-conditioner or heater.

Patterson half ran, half fast-walked, got to his car by 3:35, and made it to the exit by 3:38—a good day. A couple minutes later and he'd have been stuck in gridlock until 4:20 or later.

McCann's attempt, though, was unlucky. He power-walked through the gate, made his way up and over to Sixty-Eighth Street, cut through the gap in the trees, and ran down the hill to where his car was parked. He jumped in, started it, and hit the gravel road, trying his best to accelerate while navigating the potholes to a back road that might just get him out of the lot in time to beat the traffic jam. But he didn't make it to the exit until after 3:38, so he sat there, turned off his engine, and listened to the radio for the next hour.

Far worse commutes awaited some. For example, the yard sometimes needed workers on an existing carrier maybe a hundred miles off the coast. In that case, shipyard workers made the reverse ride through the tunnels to get to the Naval Air Station at Norfolk, and there they caught a Grumman C-2 Greyhound prop plane making carrier onboard delivery, or COD, out to the ship. Some of those flights took place in the fall, when hurricanes or other storms turned the

Atlantic to a swirling sea of contrary winds and mountains of motion. As the carrier's air boss struggled to land fighters on their last whiffs of fuel before bad weather ended all flight ops, an incoming COD flight would be put into a holding pattern, making racetrack-like circles around the carrier as its movements made world-class rollercoasters look like Disney kiddie rides. On those flights—circling, rocking, and oscillating on the COD, with the passengers seated facing the rear of the plane, with only one window in the middle of the fuselage, all on board—even seasoned sailors and officers used the air-sickness bags.

Working on a ship at sea was no picnic either. Little Ed was on the *Ford* off the Virginia coast during its first ship trials. He and his team checked out hoses on the far aft end of the flight deck, where the only thing separating a body from the end of the ship and the roiling Atlantic was the row of little metal baskets supposed to catch folks if they went rolling off the back of the ship. He stood on the edge, bracing against the wind cutting across the flight deck, topping out at about thirty-five knots, the carrier kicking up a wake of whitewater at its stern. Suddenly, the carrier made a hard starboard turn into the wind, and the carrier loudspeaker crackled. "Get below deck! Sixty-plus knots across the flight deck!"

Sixty knots! Almost seventy miles per hour! If Wingnut had wings, he would have been airborne. He didn't have time to get belowdecks, but he sure as hell hit the deck. Actually, the wind slammed him to the deck. He had no idea wind could be this powerful. He and his workmates started to crawl across the deck, all the way from the fantail to the island. It was like clinging to a bobbing pier in the middle of a hurricane. He focused on every movement, trying not to think what would happen if he lifted his body just a tad too high or if he lost his footing. He imagined his body rolling across the deck. Would that flimsy little basket really stop him from rolling right into the sea?

What's wrong with this captain? What was he thinking?

Hats, vests, and even prescription glasses had all disappeared like falling leaves in a strong autumn gust. He could not recall a scarier moment on the job, in or out of the yard.

The yard often tasked shipbuilders to work on other carriers, in and out of the yard. Those assigned to the *Kennedy* found themselves putting in more time on CVN 78 than on CVN 79 as the yard and the navy battled to fix the *Ford*'s systems. As the supervisor for both ships, Lucas Hicks got a real firsthand education on what it meant to put a new ship class full of new technologies, like its takeoff and landing systems, out to sea. "It's about putting the right people in place—the directors, and letting them do their jobs," he explained, "which is sometimes difficult for all of us."

The real test came when the ship hit those ocean waves. Water is the lifeblood of the ship, and carriers need water for just about every system, starting with the nuclear reactor. Once the *Ford* started operating at sea, pumping the water through the right systems and putting its new technologies to the test, then the yard and navy knew what work still needed to be done. A testing plan helped, but no plan could train the crews and exercise the systems to their maximums. It took being out on the ocean. Unfortunately, *Ford*'s systems—those from the navy and those from the yard—failed to perform exactly as designed. Common enough for a new ship class, but the testing of this one dragged on and on, because it took longer than they thought it would to find and fix all the issues.

Hicks and everyone else expected the *Kennedy* testing to go much easier; it always did for the second ship in a class. Still, Hicks pushed to build a little margin in the schedule for themselves. Now that they had outfitted CVN 79 more than any previous carrier, maybe they could launch the ship earlier, making it possible to test earlier. He gathered his team of experts and master shipbuilders and asked, "What do you guys think of doing something crazy with 79, like coming out of the dock earlier?" Crazy indeed. Carrier building schedules were set in stone, the sequencing timed just so. Any changes usually signaled a delay. Accelerating the schedule, launching the ship faster—no one did that.

But Hicks knew how the yard had struggled through the *Ford* systems testing. If they launched the *Kennedy* earlier, they could, in the parlance of shipbuilders, de-risk that shipboard testing aboard CVN 79. They had completed so much outfitting work on *Kennedy* that they could pivot from their normal prelaunch plan, button up some basics—like the primary water, galley, air-conditioning, and fire main systems—and put that ship into the James earlier than scheduled. The *Kennedy* team thought it could be done. They figured they could launch that ship three, maybe four, months earlier, pleasing the navy customer and buttressing the sales pitch in the Pentagon and in Congress to nail down those other carrier contracts and keep those steelworker jobs.

Meanwhile, in Dry Dock 12, as the new ship shape grew with each superlift, CVN 79 refused to sit still. The hull shifted as the steelworkers put on the tons of steel sections. Normally, ship designers counted on that settling of weight to help lock some major hydraulic systems, like the weapons elevators, into place. However, as on the *Ford*, the ship settling created major headaches with the new electromagnetic systems—particularly for those elevators.

8

WEAPONS AWAIT

SPRING 2014

To truly appreciate the hurdles and the potential angst presented by the new advanced weapons elevators, take a trip back to the spring of 2014, about a year before the yard and navy signed the *Kennedy* contract, when Rear Admiral Nasty Manazir was making those three-plus-hour drives—each way—to Newport News from his air warfare director's office in the Pentagon to check on *Ford*'s progress, especially the installation of the new takeoff and landing systems.

As he walked the waterfront those days, past some of the early dockside assemblies of CVN 79, he admired the profile of the *Ford* with the new sleek flat radar panels—big as billboards—on that tower, now sitting way back aft, making additional room on the flight deck for more jet operations. The *Ford* looked different from the Nimitz-class flattops. Manazir zipped up his brown leather flight jacket and envisioned futuristic stealthy F-35C fighters and Stingray drones zipping off the flight deck. Thinking back to his own days at war as pilot of an F-14 Tomcat over the Gulf, he pictured what it would be like to come to a screeching stop on a ship like the 79.

He followed yard supervisors to the new electromagnetic catapults and then to the landing recovery machinery to get updates on the problems and solutions. Manazir, remember, suffered continual hammering in Congressional hearings on issues with those systems, and he needed frank and immediate updates. He could afford no more surprises.

Then the shipbuilders led Manazir into the bowels of the ship, through all those carrier-construction work zones, to talk about the new advanced weapons elevators, the AWEs. After the Star Trek whiz-banginess of the new takeoff and landing systems, the air warfare director expected a ho-hum briefing.

The size of truck loading platforms, the elevators had more than twice the payload capacity of those on *Nimitz*, zipping along at a speed of two-and-a-half feet per second using electric actuators—little machines that opened hatches or otherwise moved the platform between the decks—instead of pulleys and machinery. Manazir climbed into one of the elevators. The metal doors clanged shut and motors whined. Lucas Hicks, then the deck machinery construction superintendent, yelled over the noise. "Before, we had to ramp up manually," he told Manazir. "Now we're running on complete automatic control. We're cycling on all day and we're into the test phase."

Nasty trusted and respected Hicks, a seasoned pro from a long line of expert shipbuilders. As an apprentice blacksmith in the yard, his grandfather shoed horses and forged metal, while keeping a journal of his years in the yard. Lucas read through it, surprised to see how much things had remained the same—the welding, the forging, the basic shipbuilding principles.

Lucas Hicks's father grew up in one of the apartment buildings erected to provide housing across from the yard gates for shipyard workers. As a boy, Lucas dreamed of playing pro football but reckoned on being a shipbuilder. His dad, though, talked him into going to college. Lucas, remember, attended the Merchant Marine Academy, as Jennifer Roman had done. Like her, he started at the yard in a satellite office, Building 800 on Jefferson Avenue, to focus on 688 submarines, the class that started with the USS *Los Angeles*, navy number SSN 688. Exiled to a gulag outside the yard gates, he did everything but build ships. When the yard made its little temporary foray into conversion of commercial ships to military supply ships, he saw his chance to use his merchant marine background to nab an engineering job in the North Yard. He moved on to construction of those converted ships and slid into overhaul of the *Nimitz*. After that, he became a fixture at Dry Dock 12 and never looked back, becoming the expert on integrating the new EMALS launch system on the *Ford*. That EMALS experience made him the automatic expert for the new weapons elevators, which used the same basic technology, although, as it turned out, in a much more challenging mechanical environment.

Shipbuilders wired the elevator for more electricity to use linear motors. The yard had ordered six thousand actuators, more than an order of magnitude more than previous orders for a carrier. Engineers had understood the physics of such a system for some time. Before the *Ford*, though, no one ever tried putting the technology on a carrier, which meant integrating products from different vendors that had never been integrated before, combining hardware and software never combined before, getting components to "talk together" that never before spoke the same language. Still, Manazir figured, shipbuilders could build elevators in their sleep.

The shipyard supervisors told the admiral the hardware and software meshed well with the lift design. Still, looking at this whole new setup, he felt vaguely uneasy. Through the decades, naval engineers and shipbuilders had honed the craft of carrier construction, developing a cadre of technicians to construct and operate the ship, as well as a training program to help sailors, supervisors, and contractors learn the ropes. These new elevators lacked that historical craftmanship. After all the problems he worked through on the takeoff and landing systems, doubt tugged at Manazir. Newport News managers told that they had started to work out the training system as they brought it all online.

"How do you know that's working?" Manazir asked. "That's not tribal knowledge."

No "schoolhouse" (special training center) existed for the *Ford* yet. Moving those weapons more quickly remained the long pole for the stepped-up carrier operations. Even if the ships could launch and recover aircraft more quickly, all that new technology would mean nothing if deck crews had no missiles and bombs with which to arm the fighters. After the yard wired up the elevators, it used them as operational trainers. The navy had no choice but to take the risk and train as the carrier construction continued.

Others, though, found the risks cropping up in the *Ford* construction program unacceptable. Back in his Pentagon office, where plaques warned of prohibitions against cell phones, cameras, and recording devices, Manazir and his staff prepared to wage battle with Senate staffers whose bosses threatened to cut the navy's carrier force because of the ballooning costs, the technological fiascos, and the unproven threat of long-range Chinese "carrier-killer" missiles that the Chinese military through the Chinese media said could sink carriers thousands of miles out to sea. It became increasingly difficult to justify the growing expense to build and support carriers, even as Congressman Randy Forbes's warnings about China's flattop plans turned into reality as China began building its own homemade carrier force. Black marker lined Manazir's whiteboard, listing arguments, pros, cons, and risks of the future and current carrier force.

Fast-forward to October 2015, a few months after the award of the main *Kennedy* construction contract. Manazir found himself under the glare of senators' stares as members of the Senate Armed Services Committee interrogated him and other navy and Pentagon officials about the problems plaguing the *Ford*. Of course, the new takeoff and landing systems ranked high on the agenda, as did the radar suite and combat system. Now, the weapons elevators made the list of worrisome *Ford* technologies.

Specifically, testified J. Michael Gilmore, the Pentagon chief weapons tester, those systems, including the elevators, represented "significant risks" to a successful first deployment of the carrier. "Developmental testing on CVN 78 AWEs has required substantial contractor support suggesting that the system has poor reliability. If the AWEs on CVN 78 are unreliable, it will degrade the ship's ability to conduct combat operations." The ship might not be able to defend itself in combat.

The powerful and pugnacious Senator John McCain pounced.

"The advanced arresting gear cannot recover airplanes," noted the Arizona Republican. "Advanced weapons elevators cannot lift munitions. The dual-band radar cannot integrate two radar bands. Even if everything goes according to the navy's plan, CVN 78 will be delivered with multiple systems unproven."

Manazir used the hearings to focus on the future.

When we looked at the future and the way that the threats around the world were going, we devised the Ford-class, with thirty-three-percent greater sortie generation capability, with enhanced technology and an electric capacity, and, with the EMALS and AAG, an ability to increase getting airplanes on and off the ship, and other technologies around the ship. That campaign model, sir, looking at threats around the world, is what delivered the requirements base that resulted in the *Ford* design you see today. When we stabilize those requirements, that is one aspect of stabilizing the cost and schedule.

By the middle of the decade, then, the navy knew about growing questions and risks associated with the elevators, as well as those about fulfillment of other requirements, and yet it took delivery of the *Ford* from the yard in 2017, approving its construction, about two years after the Senate Armed Services Committee hearing. Rear Admiral Brian Antonio, the navy program executive officer for carriers

at that time, realized straightaway that he needed to get up to speed on these systems, so he started taking hourlong briefings on the physics of linear motors at his Naval Sea Systems Command office in the Navy Yard on the edge of DC.

As the shipyard reps pointed out to those from DC touring the *Kennedy*, nineteen new technologies were combined to create the advanced weapons elevator system. Gone were the ropes, cables, and wires of the traditional elevator. The electromagnetic motors drove the elevator, while the actuators automatically opened and closed doors and hatches as the platforms approached and cleared the different decks, recharging the motors along the way. Each time a door or hatch closed, that seal had to be watertight to control damage on the ship and keep it from sinking during a catastrophic event. But actuators supplied to the yard had defects. Also, figuring out the software elevator controls proved tougher than initially reckoned.

A more basic problem thwarted efforts to fix the system. The yard had built the elevators with a hydraulic mind-set when they had really needed to employ an electromagnetic mind-set. The settling of the hull steel during construction, which had been so beneficial in locking the old hydraulic elevators more securely into place, actually shifted the electromagnetic elevator system out of alignment. Though slight, the misalignment was enough to throw the whole system out of whack. As a result, the elevators failed to meet the navy "fit-up" requirements for the doors and hatches. The legacy hydraulic elevators would force a seal with brute force, but the new electromagnetic systems required precision alignment to make the seal.

Manazir remained flabbergasted that elevators proved so complicated and problematic. Whatever the complications, they required a solution, and fast. The navy needed no more problems with its increasingly expensive new aircraft carrier. Without working elevators, the navy possessed not a carrier, but a wonderful little mobile airport for a bunch of fighters rendered useless in combat. They needed to fix

the elevators on the *Ford* to prove they could build them right for the *Kennedy* and the follow-on carriers. Otherwise, CVN 79 and the rest of the ships of the class faced *very* early retirement.

"What are we going to do about this?" Antonio asked the yard. The rear admiral started holding monthly meetings in February 2018 to look for answers.

The navy and the yard managed to keep a lid on the elevator problems through the first part of 2018 as the *Ford* entered what the navy calls the ship's post-shakedown availability, or PSA—basically pulling next to a pier for some maintenance, repairs, and upgrades. This maintenance PSA was planned initially to last eight months but was extended to nearly a year and a half to include, among other things, construction and installation of weapons elevators. The extra *Ford* work took its toll on the *Kennedy* program, too.

Welders, metalworkers, and other crafts workers who'd done that kind of hotwork on the elevators on the *Ford* moved over to the *JFK*, passing along their knowledge and lessons learned to save time and money. Then those seasoned workers, who taught metalworkers the ropes of elevator construction aboard CVN 79, found themselves drafted back onto CVN 78, to fix the *Ford*.

However, without the seasoned people working on the *Kennedy*, the remaining relatively inexperienced crafts workers struggled to figure things out for themselves, leading to trial and mostly error. Machinists climbed aboard the *Kennedy* only to find key equipment missing. *Ford* workers "borrowed" it for work on their vessel. Other times they entered their spaces to find that steel structures they'd finished and left there had vanished. CVN 78 workers had "borrowed" those, too.

On November 2, 2018, the news agency Bloomberg ran a news story headlined "U.S. Navy's Costliest Carrier Was Delivered without Elevators to Lift Bombs." Lawmakers threatened to halt the navy's carrier-buying plans, putting the yard's program at extreme risk.

In December, Senator James Inhofe, the Oklahoma Republican

who took over as chairman of the Senate Armed Services Committee following the death that August of John McCain, made the trek from DC to Newport News Shipbuilding, the first time he had been at the yard in about three years. He planned to find out why the navy had accepted the country's new $13 billion aircraft carrier without working weapons elevators. He wanted to know what the company and the service planned to do about what the navy secretary referred to as "our open Achilles Heel" on a ship whose total construction cost equaled one day of all federal spending.

With the *Ford*'s elevator problems, growing costs, blown deadlines, and issues with the takeoff and recovery technologies—Inhofe harbored second thoughts about the navy's plan to buy two carriers on one contract. He questioned the practicality of such a plan after the navy's acceptance of a ship of the same design with unfinished, unacceptable, and uncertified elevators that made it impossible for the crew to do their jobs for at least a year.

"I think the case for two right now is weaker because of the lack of success in getting everything working on the *Ford*," the senator told Bloomberg at the time. "If this were a first delay, I wouldn't be as concerned."

Once Inhofe returned to Washington, he took the matter up with Defense Secretary James Mattis—music to the ears of those running the carrier program at the yard. They found Mattis knowledgeable and fair and believed the defense chief wanted to give shipbuilders time and cover to work out the carrier problems and meet those *Kennedy* man-hour cuts. The unwritten and unofficially unspoken deal remained in effect—if the yard cut those man-hours on the *Kennedy* by the contractual amount, then the Pentagon would seriously consider—and the yard would likely get—a two-carrier deal. So, if Inhofe talked about this with Mattis, the yard's performance on *Kennedy*—not the elevator problems on *Ford*—promised to have a greater impact on the two-carrier deal. The Pentagon supported the idea. Buying two carriers on

one contract would save the nation billions of dollars. Program officials remained unsure how many billions but reckoned a savings of at
least $2 billion. The deal made sense from business and manufacturing
standpoints, enabling bulk-parts ordering and more assembly-line production, while increasing savviness within the workforce.

Also, Inhofe believed he and other committee members shared an
understanding with Mattis. The defense secretary knew, as the chairman did, that until the *Ford* went to sea with working elevators, the
navy possessed only ten operational carriers, not the eleven mandated
by law or the twelve lawmakers said they wanted, and nowhere near
the fifteen that the navy needed to deploy to meet the needs of the foreign combatant commanders around the globe. The plan presented by
the navy and the yard to the chairman showed all eleven elevators on
the ship as fully operational by about the summer of the coming year,
in mid-2019. That suited Inhofe, but he wanted working elevators on
Ford before considering a two-ship deal to buy CVN 80, the *Enterprise*, and the yet-to-be-named CVN 81.

The navy and the yard needed to fix the *Ford*'s elevators and repair
the public-relations fiasco for both the service and the ship—two difficult tasks. The yard continued to throw more steelworker-power at the
Ford, stealing from the *Kennedy* ranks. And the navy tapped high-caliber
firepower to quell those public concerns. James "Hondo" Geurts, assistant secretary of the navy for research, development and acquisition,
told the Senate Armed Services Committee that the yard remained on
course with the installation, testing, and certification of the elevators.
The yard looked to finish construction and installation of the eleven
elevators by the end of the maintenance overhaul, he said, but that certification activities threatened to drag out after the *Ford* left the pier.

The former top acquisition official for the US Special Operations
Command, Geurts learned to develop and buy things fast, often cutting through Pentagon red tape to get special forces their equipment
sooner than most of the services. Navy Secretary Richard Spencer

hired Hondo to bring that kind of flank-speed mentality to navy programs. Spencer showed a lot of confidence in Geurts.

Near the end of 2018, Spencer watched a poor Navy football team lose against Army with President Donald Trump at Lincoln Financial Field in South Philadelphia. They dressed warmly, the temperature on this typical winter day in Philly reaching just a few degrees above freezing. A former marine pilot, Spencer made a fortune as a financial investor. Unlike Mabus, a seasoned Mississippi politician known for his cautious Southern political manner, Spencer preferred a faster pace. He spoke at the speed of thought and wasted little time between decision and action. During the football game, the secretary and president talked about the new carrier. Should we go back to steam catapults? Trump wanted to know. That had become a relatively new thing of his, pushing for the carriers to forgo the new electromagnetic cats and put the old steam ones back in there instead. "Steam is very reliable, and the electromagnetic—I mean unfortunately, you have to be Albert Einstein to really work it properly," Trump had told US service members during a recent Thanksgiving Day call. Trump didn't know, or didn't care, but Newport News Shipbuilding *had* already studied reinstalling steam cats on later Ford-class carriers but rejected the idea as too expensive.

The elevator problems came up during a chat between the president and the navy secretary. As head of the navy, Spencer said he needed to be the one to take accountability for the ongoing elevator challenge. Spencer asked Trump to stick out his hand. "Let's do this like corporate America," he said. He shook his boss's hand, declaring, "The elevators will be ready to go when she pulls out or you can fire me." He had never been fired by anyone and had no intention to end that streak.

During that same winter, navy brass, yard officials, and Senate representatives began a kind of shuttle diplomacy of information and updates between Tidewater and the Beltway about the status of and progress on the elevators and the continued hope for a two-contract deal. The first elevator was successfully certified as being turned over to the crew in December, about the time of the infamous football-game handshake agreement. The yard and navy still hoped and planned to do the same for another five elevators by the time the *Ford* completed its maintenance PSA.

In their meetings with Senator Inhofe, navy carrier program officials stressed the progress made with the electromagnetic launch system—more than seven hundred successful cat launches. "All that's great," Inhofe told them. "But still, the elevators still don't work. I feel a little uncomfortable saying, 'Let's go ahead and let's get two [carriers], and everything is going to be fine.'"

It was more than faulty elevators that put the two-contract carrier buy in jeopardy: *Kennedy* man-hours jumped. Government reports near the end of 2018 put into doubt the yard's ability to meet the labor-saving plan it had promised for CVN 79, thereby weakening—in the eyes of some, nullifying—the argument for a two-carrier buy.

An event at the yard, though, overshadowed all those concerns, reminding everyone just how dangerous work could be on the waterfront.

An unseasonably warm winter morning greeted the yard with a northeasterly breeze off the James whispering over the waterfront. The clouds all but erased any hint of an emerging sunrise as the steelworkers prepared to start the first Thursday shift on this thirteenth day of December in 2018. At about quarter after six, James Goins, a master shipbuilder with forty-one years of experience, a sheet-metal worker, scrambled along the scaffolding inside one of the empty metal shipping containers that roll daily into the shipyard, literally by the truckload. Most stayed there for quite a while, stacked in rows on top of one another throughout the yard, used for storage, supplies, and break areas for shipbuilders, who liked to duck inside to chat, grab a bite, or make a quick cell-phone call in a somewhat quiet area. Some even napped.

Crews strung lights, hung scaffolding, and created their own little waterfront warrens. In tight areas, workers built stair towers between the stacked containers to more easily access one of the upper ones.

Goins's reason for climbing up to a top-level shipping container at that moment remains known only to him. What happened that precise moment, no one has determined for certain. No witnesses saw a thing. Despite the guardrails meant to prevent such accidents, Goins fell from the scaffold platform.

He fell only about eight-and-a-half feet, a relatively low-height mishap in a yard where workers regularly balanced themselves precariously at nosebleed levels. But that eight-and-a-half feet was more than enough. As Don Doverspike warned his workers time and time again—falls were the number-one cause of injury, and any fall promised to be deadly. Goins severely bruised his arms, shoulder, and back as he fell.

Emergency personnel rushed the master shipbuilder to the hospital for immediate surgery. He died about three weeks later, on the third day of 2019, due to complications from the injuries and subsequent surgery.

Following an investigation, the Occupational Safety and Health Administration cited the shipyard for violations for problems unrelated to the fatal accident, which they discovered while they were looking into the fall. Essentially, OSHA investigators found only half the amount of space required on the platform outside of the container, the space needed to walk in and out of the shipping box near where Goins fell. OSHA slapped the yard with a $13,260 workplace-safety-violation fine. The accident sparked an internal yard review of shipping containers, and the company fixed safety problems with another container well away from the deadly accident. OSHA, though, never determined a cause for Goins's mishap.

The yard remembers its dead, but can afford little time to mourn. As winter winds blew across Hampton Roads in early 2019, two Tidewater shipbuilder teams raced to finish the waterfront work on two of the nation's newest carriers, one team trying to fix the bedeviling elevator problems on the *Ford* and the other trying to stay on schedule yet keep man-hours down on the *Kennedy*. The stakes remained high, matching the level of scrutiny on both. Navy Secretary Spencer's job still depended on making the *Ford*'s elevators right. Meanwhile, Congressional investigators from the Government Accountability Office put the *Kennedy* under their own microscope. Once the GAO gumshoes targeted a program, they'd find some kind of dirt and report it right to Congress. Word spread that GAO had *serious* doubts that the yard would make its man-hour cuts as promised and contracted.

While Butler, Doverspike, and Carper pushed their teams to get final hull assemblies ready for the Big Blue crane lifts, GAO investigators dug through the *Kennedy* financials, armed with spreadsheets, data mining tools, and skepticism. As CVN 79 program officials admitted, "Shipbuilder cost performance remains . . . slightly below the level needed to achieve production labor hour reduction targets." The navy also warned about the yard's need to battle material shortfalls that disrupted its ability to build the ship.

Material shortfalls and disruptions aside, Mike Butler remained laser-focused on keeping the *Kennedy* on track to meet all its 2019 milestones. In just a few months, for example, Big Blue's schedule called for the crane to lift the ship's island into place. The *Kennedy* began to look more like a carrier, the shape of its hull and deck filling the dry dock, encased in scaffolding and covered with shipbuilders bracing themselves against the cold January wind. Butler's world was centered on the *Kennedy*, not the proposed two-carrier buy.

But a two-carrier buy—and the *Kennedy* man-hour reductions needed to secure it—very much occupied the thoughts of the navy brass and Newport News Shipbuilding leadership. A year had passed since Congress had asked Secretary of Defense Jim Mattis to study the pluses and minuses of a "block buy" for two Ford-class aircraft carriers. The pressure started to mount, not so much along the waterfront but in the corridors of Capitol Hill. A great deal of political engineering fed into the buying and building of an aircraft carrier. Companies, workers, and thus federal lawmakers throughout the country all maintained a keen interest in the construction of one flattop, let alone a contract for two.

The nation had made no two-carrier deals since the end of the Cold War, when the navy bought CVN 74 *Stennis* and CVN 75 *Truman* in 1988, and the possibility of such a deal for the Ford-class carriers galvanized pro-carrier lawmakers, suppliers, and lobbyists, starting in the spring of 2018, to make it all a reality. Ship-contract negotiating normally started at the staff levels and worked its way up. In this case, though, the impetus and the guiding forces for the two-carrier deals started at the top with Hondo Geurts, the assistant navy secretary, and Mike Petters, the CEO of Huntingtin Ingalls Industries, the parent company of Newport News Shipbuilding. While waterfront workers focused on fixing the elevators and cutting *Kennedy* labor hours, the upper navy and company echelons kept up the effort for the two-carrier deal.

"This deal could save a lot of money," Geurts explained. For Petters, buying two ships would allow the yard to cut down the time it took to build them singly, back-to-back. The deal would yield no time savings in the actual construction of the carriers, but Petters and the yard could retain the workers from one carrier to the other, instead of losing them and their experience as the shipbuilder usually did when it built one carrier at a time. In addition, contracting for two ships at one time meant the yard would save money by buying the steel, pipes, and other materials for two carriers at a time—a boon and guaranteed source of income for small manufacturers throughout the country and life-preserver work in a single-company town.

Geurts told everyone to be prepared to wait until at least the end of the year for such a contract. In the halls of Washington, they stepped up the pressure. Rick Giannini, chairman of the Aircraft Carrier Industrial Base Coalition lobby group, publicly claimed that the navy would save more than $2 billion on the two carriers by buying both under one contract. Besides chairing the coalition, Giannini also served as president and CEO of Milwaukee Valve Company in New Berlin, Wisconsin, a carrier valve supplier and one of those depending on carrier work.

After negotiating with the yard for half a year, the navy upped the ante in August 2018, when Secretary Spencer publicly predicted even greater possible savings for a two-carrier deal, saying the navy could cut at least $2.5 billion off the cost of one carrier at a time. The navy and the yard worked together on a proposal analysis that they planned to show Congress in the third quarter that year. The yard proposed contract savings of 6 percent, but the navy wanted to increase that to 8 percent. The man-hour cutting on *Kennedy* promised to seal the deal for Spencer. The yard had reduced *Kennedy* man-hours by 1.6 million, about 15 percent, compared to the *Ford*.

Distractions remained with the overall carrier program. Trump still publicly questioned the need for electromagnetic catapults on the new carriers. Members of Congress still publicly doubted the need for carriers in the new world order and warned that the ships might be vulnerable to modern missiles and other weapons. What sense did it make, they asked, to buy two carriers when lawmakers doubted their usefulness? Even within the Pentagon, some again recommended that the navy reduce its carrier fleet. If Spencer failed to secure a two-carrier contract at that moment, any one of these issues threatened to prevent any such deal later. Also, program bean counters calculated savings using suppliers' discounted pricing, ending at the end of the year.

At the end of November 2018, the yard reported about 16 percent fewer labor production man-hours for *Kennedy* compared to *Ford*. It wasn't 18 percent, but it was damn close. The Congressional Research Service, which advised Congress on naval matters, even floated the possibility of a three-carrier buy.

Carrier officials feared that any bad news about the *Kennedy* man-hours or *Ford* elevators would turn all the analysis into nothing more than nine months of busywork. Spencer's job was on the line. The Trump administration could decide on further funding cuts or even demand some weird whimsical carrier design. Congress could reinstate laws curtailing defense spending, as it had done during the Obama administration. Petters warned Wall Street analysts at the time that it appeared the Pentagon planned to prepare two budgets, one with a host of proposed cuts and another without them.

In the spring of 2018, as the *Ford* elevator problems raised ire in and out of the Pentagon and steelworkers pieced together the *Kennedy*, Petters and Geurts and their respective teams sweated over a two-carrier deal. These needed to be more than just spreadsheet savings; they had to be demonstrable and believable. The yard also had to prove it could make the *Ford* fixes work on *Kennedy* during construction, not later, after it was in navy hands. Meeting those man-hour requirements proved they could actually make those fixes and build the ship for less money. Still, with only two ships, one that had been a construction problem child and another half-built, the navy lacked data to prove its points. Also, the final deal-sweetening, cost-cutting measure—material sourcing for the two ships—became hampered by a short shelf life. Nevertheless, the industry and government teams worked more closely than in the past, and by June they had banged out all the details.

Now, Hondo tried to sell the honchos in the Pentagon, the White House, and the halls of Congress. A deal this big generally needed up to two years of salesmanship. He had only six months to get all the official approvals by the end of the year. Otherwise, all those bulk deals on materials, components, and materials would expire. Also, he needed those approvals in time to include the deal in the budget, slated for release during the first quarter of 2019. Finally, the yard had already started work on one of the two carriers that would be included in the deal, the *Enterprise*, CVN 80. The more work they did now, the less they'd save later, but if they missed this year altogether, the later and lesser savings would make the deal untenable.

By the end of December, Geurts had all the approvals he needed. He and his team won over the skeptics by guaranteeing a savings of more than $4 billion compared to two carriers under separate contracts—a *Reagan* and a half. They showed that, with the digital tools honed on the *Kennedy* and with this two-carrier deal, they'd get into serial production of Ford-class ships.

On the day of New Year's Eve, all Hondo Geurts needed to do was get all those approval signatures by end of business day, the official end of the government year and the end of all the great cheap prices that had been negotiated. Otherwise, all the hard work meant nothing. So Geurts set out that afternoon with a big blue folder filled with papers with yellow sticky notes jutting out like ragged teeth, marking places for signatures he needed to collect. Normally, Monday afternoons in the Pentagon would be as packed with people as a Friday night at a busy mall, but between Christmas and New Year's Day, only dribs and drabs of people scampered through the halls, making their way past closed or near-empty offices. On this day, with rain pouring down in sheets and temperatures hovering a bit above freezing, it seemed like a ghost town.

As Geurts loped around the five-sided building, past glass-encased relics of World War II and other reminders of more recent conflicts and US military might, he recalled the days when, as an air force major, he hunted these same halls likewise for brass to sign documents. Here he was again, this time as the assistant navy secretary, watching the rain pound the Pentagon pavilion in the middle of the courtyard, gathering signatures. Psyched about the deal, he embraced this yeoman's work on the last day of the year, hoping to finish his task in time to get the paperwork to Defense Under Secretary Ellen Lord, the Pentagon's acquisitions chief, before the end of the day. Lord and her staff had agreed to wait for Geurts, even as the clock clicked to 4:45 p.m. and the rain continued to soak the Pentagon grounds. Hondo's staff also remained in the office, standing by.

"I wanted to send them home," he said later, recalling that day. "They should have been home with their families. But the deal meant so much." His staff realized how much was at stake, and they stayed in the building—the only ones besides the police and Lord's staff—when the clock ticked to 5:30 and Geurts delivered that blue folder to Lord's office, all the signatures on the proper lines. She approved the deal that night, and everybody raced through the rain to the parking lots, empty except for their cars.

Geurts and the navy officially awarded Newport News the two-ship deal on January 31, 2019. As noted, the $15.2 billion contract promised to save the navy about $4 billion compared to the original cost estimates for two carriers separately. Those contracts covered the costs of the hulls and associated electrical, plumbing, and other systems needed to live on a ship and operate it. The other government-furnished equipment raised the per carrier cost to more than $11 billion per ship.

The navy continued to tout its deal to the public. As noted, major naval contracts with Newport News had been known to be canceled; no one was taking anything for granted. Geurts stressed the 16 percent reduction in man-hours from *Ford* to *Kennedy*, with promises that the new contract would mean a further reduction in man-hours of about 18 percent between CVN 79 and CVN 81, which, just before the final contract, had been named in honor of World War II sailor Doris Miller, a navy cook and the first Black recipient of the Navy Cross. Historians recognize Miller as one of the first US heroes of World War II. As the battleship *West Virginia* was sinking during the Japanese surprise attack on Pearl Harbor, he helped wounded shipmates, including the captain, to safety, then shot down several warplanes with a machine gun he was untrained for—a scene later recreated by Hollywood. He died in action two years later.

Unfortunately, the two-carrier contract award coincided with the release of the Pentagon chief program tester's annual report, which skewered the *Ford*, highlighting problems with launching and recovering aircraft, operating the ship's new radar, and moving weapons with its new elevators. The navy had yet to accept delivery of any elevators, the report noted.

A new problem arose. As one part of the navy negotiated with the yard to buy two carriers, other parts of the navy—budget cutters and certain operational departments—worked on plans to reduce the overall US carrier force and eliminate some of the flattops in the coming years and decades.

The latest plan emerging during that winter and spring of 2019 eliminated the upcoming midlife nuclear refueling for the *Truman*, to retire the ship about halfway through its planned lifespan, and then use the money to pay for a new fleet of unmanned ships. The navy, like other US military services, had become enamored with the idea of building robot boats, ships, and subs to patrol the seas and monitor enemy forces while cutting down on the number of sailors put in harm's way.

On the Newport News waterfront, some hyperventilated. Midlife nuclear refuelings and overhauls were a major part of the shipyard's carrier business. In the grand scheme of things, building *Kennedy* and other new Ford-class ships remained the preferred line of business, but loss of the refueling work, worth billions of dollars, would also mean thousands of steelworkers' jobs.

By eliminating the *Truman* refueling and mothballing the carrier earlier, the service expected to save about $6.4 billion over several years. Members of Congress rejected the idea. Any attempt to cut the required eleven-carrier force required the approval of US lawmakers, an action they had rejected previously. Pentagon and naval officials had discussed delaying or canceling carrier refuelings before, but the option had started to gain more traction among a growing number of Defense Department strategists.

All in all, shipbuilders faced continued pressure, despite the new dual-carrier contract, to fix the *Ford* and build the *Kennedy* right the first time with even less labor.

9

TRUMP CARD

Usually, the waterfront considered a ship visit by the chief of naval operations, or CNO—the number-one uniformed officer in the US Navy—an event to embrace, a moment to shine and show off. *Ford* crew members treated the February 5, 2019, visit by CNO Admiral John Richardson aboard in Newport News—just after the two-contract carrier award and just before the Pentagon made public its plans to cut the carrier force—as such a happening. His visit there showed his support for their efforts, and the crew would get to see their most senior boss, a big deal for a bunch of sailors whose average age is just above voting age. They literally rolled out the red carpet for the admiral, and sailors crammed shoulder to shoulder in the cramped ship space to hear Richardson tout the *Ford* as "a centerpiece of innovation."

Richardson made the trip from the Pentagon to Tidewater to do more than rally the *Ford*'s sailors. He went there to inspect the weapons elevators, remind the yard about the importance of fixing them, and poke the fire of urgency to finish the work. Yard supervisors, then, sounded a more somber tone than the sailors upon the arrival of the navy's highest brass on the waterfront and on the ship. It didn't help matters that wherever the CNO went, the media surely followed. The navy wanted to ensure that the yard worked the elevator problem hard and, with the secretary's job still on the line in the case of failure, the service leadership wanted to show its determination to move ahead smartly.

Reporters never knew what to expect from Richardson. Inconsistent about when and where he'd invite the press for his events, he also made sure the navy kept the media—and thus, the public—in the dark about the latest and greatest things the service intended to develop, test, and deploy. In the past, the navy had often implored the press to cover new technology, operational concept changes, and so on—in part because the service wanted the publicity to prove that no taxpayer money went wasted. Naval managers often used the press to fight funding battles for the navy and Pentagon.

However, with the recent sizable budget increase for the Pentagon, the navy, and the shipbuilding programs provided by the Trump administration and the new Congress, the navy no longer needed to make public sales pitches through the press. Indeed, with both Russia and China making significant military strides lately, the navy stopped bragging about technology. Richardson sent out a memo to navy public affairs officers muzzling them in discussing attributes of naval equipment, programs, and platforms. This marked a departure from the stance taken by previous CNOs, particularly Richardson's immediate predecessor, Admiral Jonathan Greenert—like Richardson, a career nuke submariner—who cultivated media relationships and publicly detailed the benefits of what the navy wanted to buy, build, and deploy.

Greenert, along with Secretary Mabus, kept the navy on a solid course, slow and steady. But this new Spencer-Geurts-Richardson team—they moved fast and mixed things up. They saw the competition, or even conflict, with China and Russia as something to worry about *now*, not in the *future*. They needed tomorrow's technology to-

day. They needed the *Ford* in the fleet yesterday. But the elevators simply refused to cooperate. The future of CVN 79 *Kennedy* and the rest
of the class depended upon fixing the *Ford*'s lifts.

Navy and shipyard chaperones guided Richardson through the
Ford to the open bay leading to the elevator platform. The CNO wore a
white hard hat but not the traditional white navy uniform that Greenert usually wore on ship visits. Instead, Richardson outfitted his spare
frame in green camouflage to project an image as a "warfighter" rather
than an administrator.

He strode up to the elevator opening, stopping to check out the
easels set up in front of the lift, which held charts showing all the benefits of the new elevators, complete with pretty graphics. The CNO,
whose mind retained information like a computer, knew all that data
already. The navy put those charts there for the media, to help reporters put all that information in their stories posted all over the internet
later that day.

Richardson entered the bay and climbed into the elevator. He nodded a few times as other naval officials explained their efforts to fix
the elevators. He stepped back, and the elevator operators put on a
little show for him. They started the lift. Metal clanged while motors
whirred, the sounds ringing loudly through the ship space. Richardson nodded again and strode back out to meet the press, to accomplish the real mission of the day, persuading reporters of his certainty
that contractors had solved any technical matters and fixed the elevator issues. "Now, it's just a matter of getting through installation," he
claimed. "Elevator by elevator. We're past the uncertainty part of the
elevators. We're now just getting the work done on installing them."
Addressing the Pentagon's recent report that publicly cast doubt on
the *Ford*'s ability to launch and recover aircraft at the sortie rate adver-
tised, Richardson cast his own doubt on raising such issues so early
in the ship's operational life. "To get a meaningful sortie generation

rate, the whole ship has to operate," he said. "It's far too early to make any conclusions about sortie generation rates while we're still working through startup issues."

The *Ford* continued to launch and recover aircraft. Nothing more mattered at that moment.

Maybe—but the navy failed repeatedly to publicize that good news about the ship. Thanks to Richardson's closed-mouth policy, the navy continually rejected requests to put reporters aboard to watch all those aircraft taking off and landing. That coverage—with pictures and video—often made all the difference. Instead of talking about elevators that failed to work, folks could watch videos of fighters shooting off the new carrier. Behind Richardson's back, many navy officers bemoaned the lack of positive coverage.

One of the reasons the navy anticipated the *Ford*'s deployment, Richardson said, was because of the kind of futuristic weapons the navy planned to put aboard that were not yet in current carriers' arsenals. The CNO talked about lasers to shoot down enemy missiles, microwave beams, and other directed-energy weapons. The navy wanted the *Ford, Kennedy*, and follow-on carriers to serve as electronic-warfare centers. As such, Richardson maintained, the ships promised to be even harder targets to take out, despite warnings from anti-carrier strategists. "It is the most survivable airfield within the field of fire," Richardson said. "This airfield can move seven hundred twenty miles a day. In many ways, the carrier is less vulnerable than at any time since before World War II."

While Richardson saw the carriers as less vulnerable to enemies' missiles, subs, and ships, concern grew about their increasing vulnerability to enemies within the Pentagon, the White House, and the halls of Congress. With the elevators keeping the *Ford* in the spotlight and the Trump administration putting a target on the *Truman* for early retirement, the CNO's visit took on even greater importance. So did the work on the *Kennedy*, being built in Dry Dock 12 just down the waterfront from the *Ford* as Richardson was getting his look at the lifts. Even though Newport News Shipbuilding had its two-ship deal, hitting those man-hour reductions was just as important as ever. If the yard failed to prove its ability to build a carrier with that promised manpower, the two-carrier deal could be the swan song of nuclear carriers for the company and for the navy's future carrier plan. The Pentagon at that moment put such a scenario under a microscope as it considered new carrier studies and the proposed elimination of the *Truman* refueling. Also, GAO investigators found more evidence of the yard's inability to meet its *Kennedy* man-hour-cutting numbers. It looked grim. Then the yard and the navy got a lifeline of sorts from the unlikeliest of sources.

On one of those hot and humid late spring days in Hampton Roads, Vice President Mike Pence climbed aboard the carrier *Truman* as it sat moored at the Norfolk Naval Station pier across the mouth of the Chesapeake Bay from Newport News Shipbuilding. Everyone aboard gave the VP a rousing welcome. Before he arrived, officers ordered sailors to cheer for Pence the way they did for a strip-club pole dancer. Despite the near-record heat, he stood before the sailors—a clapping sea of green in their cammies—in the hangar bay wearing a brown leather flight jacket, with a blue tie.

"It is my great honor to join you aboard America's lone warrior, the USS *Harry S Truman*, where for more than twenty years you've been 'giving 'em hell' on every deployment," Pence told the sailors, who followed their orders and applauded him with lusty emphasis.

"Last year, President Trump signed the largest investment in our national defense since the days of Ronald Reagan and called for the building of a three-hundred-fifty-ship navy. And we are on our way," Pence boasted. More spirited clapping—although more than one sailor was thinking, *Yeah, that isn't going to include our ship. You're gonna mothball our ship.* They knew the Trump budget called for an early *Truman* retirement to pay for a bunch of robot boats.

"You know, it really is an honor to be back aboard the *Truman*,"

Pence went on. "And looking out at all of you and your shining faces is incredibly inspiring. You all serve this country with great distinction. There's a lot of stories to tell about the people who have served on this ship—past, present, and future."

What future are you talking about, Jack? Not for our ship.

"For more than twenty years, the USS *Truman* and its crew have played a vital role in defending our nation and protecting our interests across the globe. And it's humbling for me to stand among all of you that have continued that tradition. This ship has served as a constant sign to the world that we will always ensure our security and we will always stand for peace through strength."

Sounds like a eulogy.

"And as I stand before you today, I know that the future of this aircraft carrier is the subject of some budget discussions in Washington, DC. And as we continue to fight Congress to make sure that our military has the resources you need to accomplish your mission, President Donald Trump asked me to deliver a message to each and every one of you on the deck of the USS *Truman*: We are keeping the best carrier in the world in the fight. We are not retiring the *Truman*."

The sailors clapped, cheered, and whistled, for Trump was going to give Congress hell and keep their ship going for decades to come. Some sailors knew the truth. The president wasn't standing up to anyone in Congress. If he was standing up to anyone, he was standing up to himself. Congress never made the call to retire *Truman* early. As shown earlier, leading lawmakers fought against such a move. The *president*'s proposed budget called for the *Truman* retirement. *President Trump officially asked Congress to approve the carrier cut*, not the other way around. Pence and Trump cut their own administration off at the knees with Pence's promise to save the *Truman*.

Presidential Pentagon budgets usually work something like this: the administration and the Defense Department ask for funds for all kinds of programs, but they also purposely leave out programs or recommend cuts for ships, planes, or something else when they know that someone on the Armed Services committees in the Senate and House will strongly object and demand that money be provided. In those cases, the administration then puts the onus back on lawmakers to find the funding—a proven negotiation gimmick.

Lawmakers had no desire to retire the *Truman* early. But with Pence's little speech, the administration talked itself right out of any serious negotiations. If Trump wanted to put *Truman* back into his own budget, great. Let him come up with the money or cut something else.

However backward, the *Truman* funding restoration showed some carrier love by the administration. Whatever warm and fuzzy feeling buoyed the shipyard evaporated the very next month. In May, GAO investigators officially questioned the estimated *Kennedy* labor-cost reductions in their public report to lawmakers: "Costs for CVN 79 are also likely to increase as a result of optimistic cost and labor targets, putting the ship at risk of exceeding its $11.4 billion cost cap." As for the promised man-hour reductions, the GAO said, "Our analysis shows the shipbuilder is not meeting this goal and is unlikely to improve performance enough to meet cost and labor targets."

The GAO also reported that the unrealized reductions will affect the promised savings on the *Enterprise* and the *Doris Miller*. Investigators called into doubt the promised lower costs. Even the navy acknowledged to the GAO that, with CVN 79 construction a bit more than half complete, material shortfall that disrupted work threatened the labor-cut promises.

In one fell swoop, the federal investigators not only cast doubt on the *Kennedy*'s savings, but also called into question all the negotiating spearheaded by Geurts and Petters. Essentially, the savings being promised for the next three carriers were little more than a mirage, the GAO claimed.

The shipyard and the navy responded to the GAO report with another major superlift by Big Blue—to give *Kennedy* that special little something to mark its silhouette forever as an American supercarrier: its island. Once planted aboard, it would rise seventy-two feet above the flight deck, occupying fifty-six feet of the length of the ship and thirty-three feet of the width, mostly on the starboard side. The towering monument of metal was to serve as *Kennedy*'s eyes, ears, and brain for about half a century, a mini skyscraper planted on the deck of the warship.

May 29, 2019, promised to be another of those sultry, sunny, sweaty summer days on the James waterfront as riggers hooked up Big Blue to the massive island. The superlift attracted attention. Navy brass and yard executives attended, along with media, and VIP guests. The big ice-packed barrels filled with water bottles ran dry early while folks searched in vain for something to drink and a little shade on the waterfront living up to its old Hell's Half Acre moniker. Mike Butler wore two hats at such happenings—on one hand keeping an eye on the shipbuilders and the job at hand, and on the other being one of the shipyard hosts for brass, media, and guests to move them where they wanted or needed to be or to answer any questions about the ship, the island, or the day's operation. In the nanoseconds between, he permitted himself to enjoy the moment. Something about this, about the tradition of the island lift, touched the very core of his shipbuilder's soul. This tradition traced its roots to the earliest days of the construction of sailing ships.

No aircraft carriers or carrier islands existed centuries ago. But Newport News steelworkers lifted and placed the carrier island in much the same way that previous shipbuilders set the ship's mast—called stepping the mast—for hundreds of years. Indeed, the day's festivities featured a scene mirroring a ceremony dating back to those times when builders placed coins at the base of a mast of a ship under construction to bring good fortune. In the case of the *Kennedy*, plenty of fortune coinage satisfied the tradition. Military brass use coins like modern-day calling cards. Every commander, office, and program has a dedicated coin, and bequeathing one of them to another involves a

Newport News Shipbuilding dominates the city and the region.

While Newport News Shipbuilding has become increasingly reliant on computer-aided construction processes, the yard still depends on its experienced labor to build the nation's massive carriers.

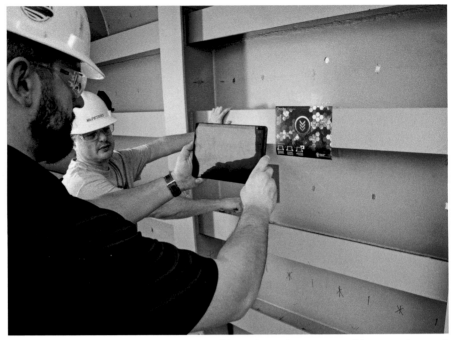

Steelworkers at Newport News Shipbuilding depend more on tablets as the yard moves into digital operations to adapt to manufacturing and ship-construction technology to meet navy schedules and funding constraints.

While his members built the carrier *John F. Kennedy* during a deadly pandemic, Newport News Shipbuilding Steelworkers Local 8888 president Charles Spivey oversaw one of the union's most successful membership drives.

Tasked by shipyard executives to find a whole new way of constructing the carrier *Kennedy*, Mike Butler pushed his team to complete larger sections of the ship earlier in the building process, saving time, money, and manpower.

Captain Todd Marzano—call sign Cherry—sought advice and guidance as the new carrier *Kennedy*'s first commanding officer from the retired admiral Buddy Yates, the first commanding officer of the first US Navy carrier USS *John F. Kennedy*.

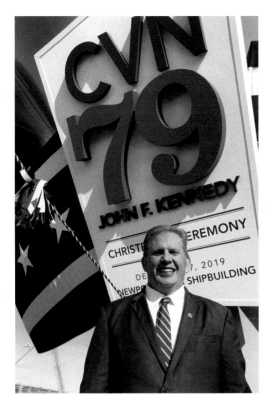

Rear Admiral Michael Manazir, call sign Nasty, grew annoyed and frustrated with frequent requests for additional funding to pay for new systems designed for the USS *Gerald R. Ford*. The cost overruns on *Ford* forced the yard to promise to cut the manpower needed to build the next carrier, the *Kennedy*, by nearly a fifth.

James "Hondo" Geurts, the assistant secretary of the navy responsible for acquisitions, oversaw the navy's work on *Kennedy* and spearheaded the agreement to buy the next two carriers in a single deal, guaranteeing work at the yard for another two decades and securing tens of thousands of steelworker jobs.

right: The *Kennedy* is a Ford-class carrier, which features twenty-three new technologies over the previous class of US Navy carriers, making the ship more lethal, effective—and challenging to build.

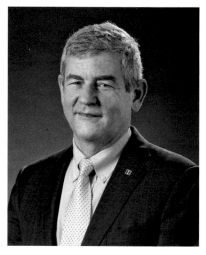

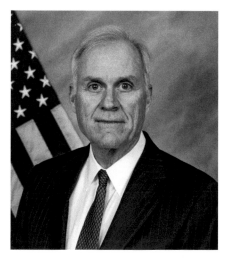

Michael Petters rose from being a US naval submarine officer to be the CEO of the US Navy's largest shipbuilding company, constructing the nation's entire fleet of nuclear-powered aircraft carriers.

The pressure to make operational the weapons elevators on the nation's new carriers, including the *Kennedy*, was so great that Navy Secretary Richard Spencer told President Donald Trump to fire him if he failed to do so.

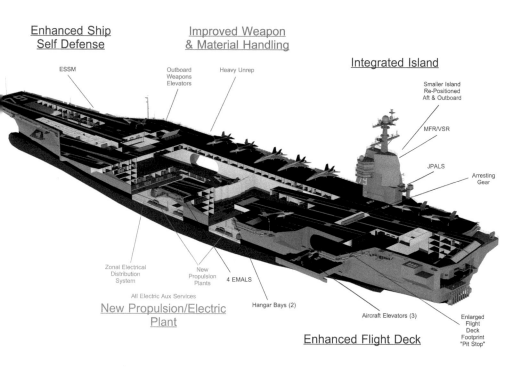

Enhanced Ship
Self Defense

Improved Weapon
& Material Handling

Integrated Island

ESSM

Outboard
Weapons
Elevators

Heavy Unrep

Smaller Island
Re-Positioned
Aft & Outboard

MFR/VSR

JPALS

Arresting
Gear

Zonal Electrical
Distribution
System

New
Propulsion
Plants

4 EMALS

All Electric Aux Services

New Propulsion/Electric
Plant

Hangar Bays (2)

Aircraft Elevators (3)

Enlarged
Flight
Deck
Footprint
"Pit Stop"

Enhanced Flight Deck

The yard builds carriers from the bottom up and from the middle out, fitting into place large sections of the ship called superlifts.

President Barack Obama visited the yard to promote his defense budgets. His presidency would highlight divisions in the region and the yard workforce.

President Donald Trump visited Newport News Shipbuilding and promised to support a big carrier-driven fleet—even though his administration was planning to make cuts in the numbers of the giant warships. His presidency further eroded relationships in the yard workforce.

Newport News Shipbuilding steelworkers watch the yard's Big Blue gantry crane mate the bow of the aircraft carrier *John F. Kennedy* with the rest of the carrier hull, marking one of the later milestones in the warship's construction.

From his perch in the operator's seat of the giant gantry crane Big Blue—the largest in the hemisphere—Charlie Holloway lifts ship sections the size of buildings weighing hundreds of tons.

Steelworkers attach *Kennedy*'s giant propellers just before launching the ship—a task fraught with risk.

Steelworkers, sailors, and others involved in the building of *Kennedy* check below the hull before the dry dock is flooded and the ship is launched.

Lee Murphy came from a family of shipbuilders. After more than three decades working at the yard, he oversaw the pierside final assembly platen work for the carrier *Kennedy*. Shortly before retiring, he gave the final order for Big Blue to lift the ship's island.

Big Blue prepares to set carrier *Kennedy*'s island on the ship deck, marking another major milestone in the vessel's construction.

Before Big Blue lowered *Kennedy*'s island to its deck, navy and shipyard officials placed coins and other important mementos on that mark, to embed them into the ship permanently.

The "plank owners"—the first sailors assigned to the carrier *Kennedy* while it was still being built—gather behind commanding officer Captain Todd Marzano while he cuts the ribbon for the first building at the yard designated as the land base for the new ship.

The US Navy needed *Kennedy* shipbuilders from the yard on the USS *Gerald R. Ford* while the *Ford* operated at sea to fix problem-plagued elevators and other systems on the operational carrier. The yard workers used that experience when they returned to the *Kennedy*.

Admiral John Richardson touted the capabilities of *Ford*, *Kennedy*, and other new carriers, but refused to allow the media access to at-sea *Ford* operations to highlight the carrier's accomplishments and offset negative publicity about the ship.

Only nine years old, Caroline Kennedy was the youngest ever to sponsor a ship built at Newport News Shipbuilding, christening the first aircraft carrier named the *John F. Kennedy* (CV 67) in May 1967, on the Tidewater waterfront.

Newport News Shipbuilding president Jennifer Boykin shows Caroline Kennedy the sweet spot to smash the champagne bottle for the carrier *John F. Kennedy*. It was not the first such moment for either.

After Covid cases started in Newport News Shipbuilding, the yard started taking steps to protect the workforce, mandating masks, social distancing, and temperature checks at the gates. The pandemic wreaked havoc with shipyard shifts, supply chains, and shipbuilding schedules.

To prove the *Ford*, *Kennedy*, and later carriers are combat ready, even with their new systems, the *Ford* underwent shock trials in the summer of 2021, during which the US Navy exploded underwater charges with the force of 3.9 magnitude earthquakes.

It takes about ten hours to flood the dry dock with 160 million gallons of James River water—enough to fill 242 Olympic-size swimming pools.

Featuring a new radar suite, the *John F. Kennedy* will be the first newly constructed aircraft carrier built to accommodate the US Navy's stealthy F-35C Joint Strike Fighters.

After acquiring an old Soviet Union aircraft carrier, renamed the *Liaoning*, China has begun to build its fleet of the ships to match and overtake the American force in the Western Pacific. *Kennedy* and other Ford-class carriers are designed to meet those Chinese threats.

Launching from and landing on flight decks near or in the Pacific and the Middle East came more naturally to the former fighter pilot than overseeing the construction of a flight deck on a ship tied up to the pier near the mouth of Chesapeake Bay just beyond the reach of the Atlantic. But this job was the path to a command of his own, something all career naval officers coveted. For a naval aviator, that path led directly to a carrier's CO chair. In the US Navy, all aircraft carrier commanders must be naval aviators.

While placing Caroline Kennedy's silver half-dollar on the deck of the ship, Captain Marzano set down his own special keepsake as well—his aviator wings of gold, the original wings he'd received back in 1995 when he graduated from flight training, not a backup set he purchased later. Doing this, he mated his wings with his first carrier command, a once-in-a-lifetime opportunity. In a real sense, he intertwined himself with the DNA of the carrier. Remember that young executive officer on the carrier *Lincoln* who, just a few years earlier, dreamed about the lucky first commanding officer for the hunks of steel then sitting around Dry Dock 12, which became the *Kennedy*? That officer was Marzano. And here he stood, that *lucky* officer. Well, he had not officially taken command yet, but the navy had selected him for the job, a little bureaucratic half-step. As shipbuilders readied the island for the lift, Marzano thought back to the summer's day in 2015.

Altogether, the coins placed down on the deck for the island-landing numbered seven, aligned in a row on a chalk line in front of a small square piece of red carpet placed in front of the point where a hatch just aft of the large 79—the numerals outlined by scores of small white bulbs, like strings of Christmas lights—on the side of the tower would rest, once Big Blue lowered it onto the deck.

With the island rigged for the lift, *Kennedy* shipbuilder-supervisor Lee Murphy clicked on the mic connecting him to crane operator Charlie Holloway sitting in his cooled-cocoon Big Blue perch above the dock and gave the order to start the tower move. Holloway tapped bit of choreography. It's done during an animated handshake. recipients usually display their wares, often on a desk or an office

Geurts offered up one of his coins, as did Rear Admiral Rc Kelley, commander of Naval Air Force Atlantic, and, of course, R Admiral Antonio, who placed two coins under the CVN 79 island his personal command coin and a navy carriers program executiv officer official coin. It meant more to him and those in his office tha just a simple milestone. Granted, the island landing would take a back seat to *Kennedy*'s christening, scheduled for later that year, or com- missioning, scheduled for early the next decade. But the island made the ship actually look like a carrier; it showed real progress.

On the civilian side, Jennifer Boykin placed her Newport News Shipbuilding president's coin, and Caroline Kennedy, John F. Ken- nedy's daughter, the ship's sponsor, provided a 1964 silver Kennedy half-dollar, one of the coins minted in 1964 after Congress voted to commemorate the assassinated president and retire the Benjamin Franklin half-dollars. She wrote in a letter for the event, "I know how proud my father would be of the ship that will bear his name and the patriotism and dedication of all who sail in her." She was unable to make it that day, so her coin placement honor fell to Captain Todd Marzano, recently selected as the first commanding officer of CVN 79.

Marzano flew an F-18 Hornet under the call sign Cherry before be- coming a ship commander. He joined the navy to fly fighter jets like Tom Cruise as "Maverick" in *Top Gun*. While Cherry possessed the same general physique as Maverick, a perfect body type to cram into a tight cockpit, a few minutes with Captain Marzano revealed a navy officer with none of that over-the-top bluster of the movie-screen Lieu- tenant Mitchell. The soon-to-be *Kennedy* CO stood straight as a mast near the carrier deck, surveying his future ship with a quiet stoicism that belied the desire burning within to assume command, put his per- sonal stamp on the carrier's culture, and guide the ship for the rest of its life.

his screen, closed his hand around the joystick, and Big Blue whirred into life. The nine cables connecting the tower to the crane tensed and strained. The island—dressed in its lift-day best with red, white, and blue bunting and an American flag waving from a pole on top—started to rise slowly off the ground like some huge alien spacecraft taking off. Then Holloway pulled it to the top of Big Blue, just below the NEWPORT NEWS SHIPBUILDING lettering, and the tower seemed to float across the sky, oh so slowly. As machinist Aaron McCann described it before, it sailed overhead like a flying metal house. Those standing on deck tilted their heads, craned their necks, and eyed the piping, cabling, and light fixtures in the sea-green compartments inside the bottom of the tower. Like the rest of the ship, shipbuilders had already outfitted the island before its superlift.

The island's flight took a while, plenty of time to chat about the day's events. Reporters there wanted to ask Geurts what the navy planned to do in the wake of President Trump's recent comments—again repeating his preference that carriers return to steam catapults. *Kennedy* shipbuilding superintendent Lucas Hicks considered himself an EMALS expert, and he remained certain that the new system gave the navy the best technology for its catapults and the future force. The navy paid the bills and if the navy decided, whether at the president's behest or for some other reason, to change back to steam—or to use whatever—then it fell to him to find a way to do it. He did, however, take exception to President Trump's comment that the new electromagnetic launchers needed an Einstein to operate them. What about the nuclear plant? Talk about needing smart people for operations. Still, he didn't want to debate the president of the United States.

Reporters joked that the navy should design a brand-new carrier, the USS *Trump*—with coal-powered boilers, hand-operated pulleys and levers operating the elevators, and medieval catapults to fling aircraft out to the sky. Geurts's aide declined to comment on that.

Those gathered at the yard for the island lift sipped from water bottles and wiped their brows in the heat. Some grabbed ice to rub across their foreheads. Navy officers dressed in their tan short-sleeves while steelworkers and supervisors wore NNS polo shirts or T-shirts. Some wore their blue long-sleeve coveralls.

While Geurts refused to address the Trump catapult plan, he did take the time to rebut the recent GAO report doubting the *Kennedy's* ability to meet man-hour reductions as planned. Speaking with a little added volume to be heard over Big Blue and the teams of steelworkers on *Kennedy's* deck, he renewed his argument that labor costs for the ship thus far ran *about* 18 percent below those of *Ford*, and CVN 80, the *Enterprise*, whose bits and pieces were already being assembled in other parts of the yard, promised even greater labor-cost reductions. CVN 79, he boasted, stood ready to launch about three months early, in November 2019.

Geurts's figures later became a bone of contention around Washington, as some argued the major reason for *Kennedy's* lower labor hours, compared to *Ford*, was because all the problems on the latter required a great deal of manpower transferred from the former, not due to more efficient waterfront work.

That debate came later, however. On that blazing late-spring day, everyone focused on the island-landing milestone. When Holloway had maneuvered the island straight up over target, Big Blue started to lower it, a relatively easy lift for the seasoned crane operator, with no flips or other fancy maneuvers. Steelworkers outlined the island-placement footprint on the deck with chalk, paint, and tape. A handful of parallel green rails on the deck also marked off the general placement of the tower. As Holloway lowered the island centimeter by centimeter, steelworkers circled and crabcrawled near the landing area, trusting Big Blue, its cables, and Holloway while they took measurement after measurement—sometimes with yardsticks, other times with eyeballs and other body parts—whistling as they worked.

They blew traffic-cop whistles in short, rapid bursts, and other times with longer emphatic tones to let everyone know what was happening with the hundreds of tons of steel overhead.

As the tower descended closer to the row of coins, Big Blue stopped, the island hovering about a yard above the deck. Workers held on to the side of the island and leaned their bodies in to get a look inside. They checked the lines on the deck. They eyeballed the space between deck and tower. They again pulled out yardsticks—only a few feet now—and rechecked and double-rechecked again. More whistling. Holloway restarted Big Blue. A centimeter at a time. Eight hundred twenty tons suspended about a foot and a half above the *Kennedy* deck. It seemed like some grand magician's trick, an impossible feat according to the laws of natural physics. It took a good chunk of the midday before *Kennedy* received its proper crown. Not a bad way to celebrate John F. Kennedy's 102nd birthday.

For Butler, Doverspike, and other shipbuilders, like the Murphys, the island lift meant more than another major milestone. The *Kennedy* now displayed that iconic carrier shape. A major portion of ship remained to build, and, as Geurts had emphasized that day, the yard, the workers, and the navy faced another major milestone coming up that fall, the christening, the biggest one yet, and they all felt stressed under the pressure. Steelworkers planned to install the ship's massive propellers that August and then paint the ship's hull and its more than four hundred tanks.

One person who no doubt looked forward to the *Kennedy* launch and christening was Tim Ewing, a supervisor working just a few piers down the waterfront on the carrier *Washington*, which recently started its midlife nuclear overhaul. At fifty-eight, Ewing had worked at the yard for thirty-nine years—one more and he earned the coveted honor of master shipbuilder. He loved his job and sharing the work he did with the Boy Scouts. An Eagle Scout himself, he had been serving as the liaison between the shipyard and the Scouts and coordinated Scout volunteers for ship-christening events like the one coming up that fall for the *Kennedy*.

On Monday, August 29, another blistering day even by Tidewater standards, when the temperature reached well into the 90s and the heat index made it feel like steel hell on the waterfront, Ewing conducted quality-control checks on *Washington*'s tanks when he slipped, fell, and died, despite vigorous rescue efforts by yard and navy.

Boykin made grief counselors available at the shipyard all week following his death. The shipyard sounded its whistle that Friday for thirty-nine seconds to commemorate his years of service. The yard later agreed to pay a $105,000 settlement brokered by OSHA, after the agency cited the company for violations, with fines reaching as high as $162,000. The yard withdrew from the OSHA voluntary protection program, removing the company from its STAR safety status with the government agency, and opening the waterfront up to greater OSHA scrutiny from then on. Boykin released a statement saying the yard had "work to do to reestablish the safety culture and program our employees deserve," and it started to make safety improvements wherever necessary.

At the union hall, steelworkers talked among themselves about how OSHA stopped even setting foot in the yard because of that STAR designation that the company *and they* had earned. Another reason they believed OSHA declined to come through the yard gates was because workers rarely actually died *on the job inside the shipyard;* most died at the hospital later. No one ever said so directly to the yard leadership, for they feared making such allegations risked their jobs. Privately, though, they made other observations. While yard supervisors made a fuss about putting safety first, foremen sometimes put trades workers' safety at risk to meet schedule. Other times, foremen made their employees work alone in unsafe spaces. That's exactly what supervisors directed them to do on the *Kennedy* to finish the ship enough to launch it that fall.

While the steelworkers readied the *Kennedy* and continued to fix the *Ford* with *Kennedy* workers, the shipyard hosted Washington royalty, intent on proving the waterfront and workforce were worth those big carrier contracts. Senator Mazie K. Hirono, the Hawaiian Democrat who served as ranking member of the Senate Armed Services' Subcommittee on Seapower, which oversaw all navy shipbuilding programs, visited the yard and the Apprentice School on September 20 and released a statement lauding the yard. "I saw firsthand the ingenuity and dedication of the shipyard's workforce and will continue to be a strong advocate in the Senate for our shipbuilders."

Just the tune the Tidewater waterfront wanted to hear.

Five days later, Defense Secretary Mark Esper visited the yard and the *Ford*, the Pentagon chief's first visit to Newport News Shipbuilding since taking the job that summer. No one knew at the yard then that Esper's staff back in the Beltway had forged a plan to cut the nation's future large-deck carrier force.

10

THE FLOOD COMETH

FALL 2019

With the *Kennedy* christening and launch set for December, the yard painted the hull and installed the giant propellers and shafts in September, with no margin for delay. Shipbuilders first blasted the hull, a dirty, dangerous, and miserable job that the yard hired outside contractors to do. In the union hall, steelworkers made it clear they wanted none of that work. The complex painting involved five different kinds of coats—and not anything available at Home Depot. Painters had to wait for the perfect temperature, humidity, and other weather conditions.

In other words, exactly *not* the kind of operation Mike Butler and his teams wanted to tackle with a hurricane bearing down on Tidewater, as they did during that first week of September 2019, with Dorian set to make landfall on the Virginia coast on September 5 as a tropical storm, right in the middle of painting the ship. Like the rest of Hampton Roads, Newport News Shipbuilding had survived plenty of rough weather. The whole region learned how to batten down the hatches. Old-timers still told stories of the hurricane in August of thirty-three that shut down the yard for a few days. Dorian appeared far tamer, but thanks to the recent rise in sea level in Tidewater, any storm, even one that promised mostly rain, caused greater flooding concerns than in decades past. The Newport News shipyard shut down everything it could and tied down everything else while, across the water, the US Second Fleet in Norfolk ordered all its ships and aircraft to prepare to scoot within twenty-four hours, depending on the hurricane's path and strength, already a source of some confusion. When the storm first struck the US the week before in the Gulf, President Trump augmented the hurricane's path with a Sharpie-drawn line on a weather map.

Hampton Roads first felt Dorian's dander later that afternoon. It grew more dangerous overnight and into Friday, dumping three to six inches of rain with gusting winds on Friday reaching fifty mph, snapping and uprooting trees, hurling debris, and cutting power. At high tide about four o'clock Friday afternoon, the most serious flooding made a mess of Tidewater. Water rose to seven feet, about the flood level during Hurricane Sandy in 2012. Other high tides caused minor flooding through Saturday.

Dorian weakened from a Category 2 hurricane to a tropical storm, but it still whipped plenty of wind and water during the storm surges, causing a continuing surge of people all over the Peninsula to any hardware store, supermarket, or drugstore for any kind of supplies to survive once the roads closed, the power blinked out, and trees crashed down. Generators, chainsaws, bread, milk, bottled water, canned soup, flashlights—the lines got long and tempers short, though most folks remained civil. Some hammered boards over their windows, and others filled up their fuel tanks and headed off to a relative's house inland, while quite a few predicted nothing more than a bit of a bad blow. In Newport News, city public works stormwater crews drove around in neon vests and hard hats, inspecting the storm drain system and clearing debris. Other city workers inspected and fueled generators they needed later for traffic lights, pump stations, city buildings, and water treatment plants. They readied front-end loaders, excavators, and other equipment needed to clean up after the storm. Newport News

Waterworks workers hunkered down for the duration to monitor reservoir levels. Construction workers tied up everything at their sites. Television and radio stations chattered with alerts. The city refreshed its website by the minute and set up a special contact-center number.

Down the road in Hampton, the city manager's office strongly encouraged those who lived in the low-lying zones to evacuate to higher ground. The city planned to open two shelters that evening on Big Bethel Road, equipped to handle those with medical needs. People with pets slogged to nearby Bethel High School. People evacuating to shelters followed instructions and brought bedding, toys for children, medicine, and any special food items, but no weapons or alcohol. Residents who needed to leave a car on high ground used the city's only downtown parking garage. Area schools canceled extracurricular activities Thursday afternoon. Langley Air Force Base closed early that day and opened only to essential personnel through Friday. Some braved a rising tide and blustery wind gusts early Thursday afternoon at Buckroe Beach on Chesapeake Bay in Hampton to get a fast dip and bounce in the roiling surf. With an average depth of twenty-five feet, and most areas running fifteen feet or shallower, such strong waves were rare. Watermen's boats chugged to shelter.

Back at the yard, CVN 79 shipbuilders readied *Kennedy* to ride out the wind and wet weather. The hull painting delayed, workers inside the ship closed openings where necessary, racing to cover any bare metal and doing whatever else they needed to do before the forced break. When it came to carriers, painting took on a whole new magnitude, as an art, a craft, and a science. Imagine painting a whole city of steel to withstand sea, salt, wind, heat, and cold, not to mention a nonstop battering ram of wave after thunderous wave. And that's just the outside; inside, hundreds of tanks, thousands of compartments, and all the spaces in between all needed a specific paint.

Anyone involved in carrier-painting, from the brush wielders to the spreadsheet-checking supervisors, became an expert weather forecaster. The Weather Channel could have cited some of the daily waterfront planning briefs, especially when some of the main painting events were approaching. Rain, heat, humidity—even the dew point—all checked whatever affected the painting schedule or process. Workers joked that *Kennedy* construction supervisor Don Doverspike had become an amateur meteorologist by the time they were ready to paint the ship. While some prayed for rain to save their grass, those working a carrier prayed against rain to save the painting schedule. And the painters suffered through a devil of a time in the Tidewater summer with the daily weather forecast refrain of "hazy, hot, and humid with a chance of afternoon thunderstorms." They all welcomed the fall, when the weather turned more paint-friendly. Except when hurricanes threatened the coast. Doverspike watched Dorian's approach with foreboding. The threat or thought of thunderstorms kept him up at night when he oversaw the painting work on the *Ford*. One storm threatened to delay the paint job for three days. Imagine what a hurricane could do.

Yard workers, though, recognized Sam Carper as the real carrier-painting guru. During the previous decade, he'd shepherded the waterfront to the more durable high-solids paint for the hull, a plus for the original shipbuilding and later maintenance. The change represented more than switching from flat white to eggshell. It was more like going from wood to aluminum siding on a house, and it meant creating a whole new way to paint the ship. Previously, the steelworkers sprayed the paint straight from paint guns, mixing the solvent with the paint as they sprayed it. But the new paint dried so fast that the sprayers had to run alongside the ship as they sprayed to keep the flow going. If they stopped for an instant, the paint dried at the tip of the gun and clogged it up.

Carper earned his paint reputation by ensuring the correct painting of all the bilges during the Nimitz-class overhauls. So, in 2012, the yard bosses promoted him to superintendent on CVN 78 *Ford*. They lagged behind on the ship-construction schedule, and they needed their paint whisperer to get them back on track and finish the dry-dock painting work in time for the *Ford*'s launch. He prepped and painted 622 tanks with up to four hundred painters while coordinating with anyone doing hotwork at the same time—a monumental task. The tank paint jobs on *Ford* turned out to be even more arduous than they were on the Nimitz carriers. The new carrier design moved all the piping and other innards from the bilge area into the tanks, to make it easier to move along and through the bilge areas, but the change was now forcing painters to squeeze into spaces with a ceiling of about four feet and climb under, over, or around all the pipes while carrying spray hoses, blast hoses, and other painting gear—and then pretzel up their bodies somehow in these spaces for an eight-hour shift of hard, suffocating, claustrophobic work. The tanks, thousands of square feet of them, spanned the entire bottom of the carrier.

Once painters wiggled their way into the tanks, they wrapped the pipes to protect them as workers blasted away all the preconstruction primer from the steel. Painters then unwrapped the pipes and did a second, less intense round of blasting to rough up the surfaces, taking care that their blast nozzles didn't punch holes in the piping, which would lead to frowns, reprimands, and rework. With all the blasting done, the painters sprayed the entire space, piping and all, their bodies pressed and curled up against the side of the tank, spraying as carefully but as quickly as possible with the high-solids paint.

Some jobs, Jordan Patterson joked, he did ass-backwards. He had to back his butt in first just to cram himself properly in the space. He appreciated his stretching exercises those days. For other jobs, he found himself upside down, hanging from a pipe by his feet like a literal Batman, one foot on a pipe and the other jammed into a bulkhead, like a human triangle. Forget the hard hat; it just fell off. Often, with no adequate ventilation, he almost choked from the fumes, the taste of the paint-solvent mix sticking to the roof of his mouth.

Painting the bilge areas carried its own set of worries. For touch-up work, deckplate steelworkers mixed the solvent and paint in a bucket, tied the handle to a string, and lowered it down to the painters below. First, they tried plastic buckets, but the chemical reaction produced too much heat for the plastic, so they switched to metal buckets.

By the time *Ford* launched, Carper's teams had finished more tanks than had been done on any previous carrier. And shipbuilders needed a similar performance on the *Kennedy*. But first they had to survive Dorian.

Butler climbed up the ramps and scaffolding and entered the *Kennedy* to make sure workers had properly tagged and stowed everything, especially as the day's weather report became more ominous: the storm was going to rake them with much higher winds than they had initially planned for. Oddly, inside the ship, with the HVAC and other fans running full force, it actually *sounded* like a hurricane, or like the roar of a jet engine constantly on full power. Many compartments looked like tornado-destruction zones, with piping, vents, and metal rods or other pieces strewn all about. Wires and cables hung from ceilings and light fixtures like jungle vines. Orange ladders, now empty, stood at attention like sentries. Butler burrowed through and made his way to the thirteen-thousand-volt electrical distribution room, the central power control for the ship, run by electrical foreman Wesby, whose electrical distribution software program, a by-product of digital engineering, made it easier to prepare the *Kennedy* power clampdown for Dorian.

That program paid off that day. No longer did Wesby's electricians need to de-energize—cut power—for an entire segment of the network for issues like overpopulated or shorted wiring. Workers used the programming to check it digitally first and then pinpoint exactly where they needed to direct their effort, eliminating rework. Wesby and his team were already a million feet of cabling ahead of schedule.

From Wesby's power control center, Butler moved on to the ship's shiny new medical compartment. Now this shone like a carrier ship space, with nearly finished operating rooms, post-op room, and doctor's office. Only the final outfitting and painting needed to be done.

Butler then made his way up the metal stairway inside toward the top of the *Kennedy* island. He entered a hatchway near the very top and anchored himself on the port side of the bridge, right behind the spot marked off for the later installation of the captain's chair. Looking like a seasoned shipbuilder in his jeans, blue plaid shirt, and hard hat, he peered out from his safety glasses over the construction zone that marked the flight deck. To his right, steelworkers secured cables against the green bulkheads and taped large X's across the big rectangular windows that wrapped around the bridge. From his perch, Butler eyed the James River Bridge, the brick-and-metal fab building, and the now-still cranes along the waterfront. The dark storm clouds muted the scene into a sepia landscape. Gusts of wind rattled the windows.

This inconvenient hurricane smashed into the *Kennedy* dry dock thirty-nine days before the scheduled launch. On the dock outside, Jordan Patterson and other painters locked and secured their tools. The ship's schedule called for the steelworkers to paint those six-hundred-plus tanks inside the ship. Dorian would cost them precious time.

As Butler walked along the side of the long dock, the carrier rose on his right, the maroon-colored layer close to the keel topped by a solid ribbon of black paint, and above that, the warship-gray hull rose up and angled out over the heads of the steelworkers scrambling alongside. A curtain of a dozen cables, each thicker than a steelworker's leg, stretched from the bottom of the dry dock to above the flight deck between two scaffold towers. Down near the bottom of the dry dock, on a smaller scaffold fort, two hooded steelworkers, one in a blue suit and another in red, worked frantically on the bulging part of the hull that housed the ship's propeller shaft, called the fairwater. Shipbuilders waited until shortly before ship launch before doing the fairwater work. The shafts stretched back to the engine room 450 feet forward, and the shaft alignment needed to be precise, the main gear being aligned to the nearest thousandth of an inch. Builders did the main work at night, to avoid the sun-related heat movement. The settling of the tens of thousands of carrier tons on the 450 pedestals of wood and concrete supporting the ship in the dry dock also made the hull bend or twist, ever so slightly—but enough to potentially knock that alignment out of whack. Four shafts, four propellers, four times the worry.

Shortly before Dorian hit, Butler walked in the dry dock with Captain Cherry Marzano. "Can I walk under the ship?" Marzano asked. He wanted to take his first walk under the keel of the ship he now commanded.

Butler lived for moments like this, to show off shipbuilders' handiwork to members of the family—the yard family and the navy family. Marzano walked slowly, reverently, looking up, trying to fathom the tens of thousands of tons over his head, balanced on wood-and-concrete pedestals. Being under a carrier like that paralyzed you, made it impossible to breathe, and knotted you up in the gut and other places. Staring up at the ship's spine, Marzano thought about all he needed to do to prepare its spirit—the hundreds of ship's "plank owners," the vessel's first sailors, who would bring the carrier to life.

Marzano now needed to establish the culture of the ship, the ethos of the sailors, through the command climate he created and baked into the steel of the ship. He viewed ship culture as one huge tank that needed to be filled and sometimes repaired. Every day he went home, his wife and daughter would ask him, "Did you pour water into the tank today?"

The fifty-year-old naval officer had been prepping his entire career for such a moment, even before he served as the *Lincoln* executive officer who had stopped in his car to watch the steel assemblies piling up for the *Kennedy* and dreamed of being the CO of the new carrier. And as soon as the brass offered him the job, he sought advice from Rear Admiral John Meier, the *Ford*'s first commanding officer, who coined the phrase about the proper ship culture being "baked into the steel" of a ship. Cherry also checked in with Admiral Yates, the first CO of the first carrier USS *John F. Kennedy* (CV 67), who also gave an idea of what to expect as the initial commanding officer of this new class of ship.

As Marzano awaited his selection as the new commanding officer for the second carrier *Kennedy*, Admiral Buddy Yates invited Cherry out to his home in Virginia Beach for lunch. Wanting to impress the navy legend, the CO-in-waiting wore his best-pressed summer whites for the occasion—and then spent the meal trying not to get any of the BBQ sandwich drippings on his uniform. Yates, remember, gave the superlift order at the second carrier *Kennedy*'s keel-laying, and his voice still carried the command presence he possessed when he captained the first *Kennedy* off the Vietnam coast during the US war there. After narrating through a living room of photos spanning his career and American history and settling back into his rocking chair—a replica of the one President John F. Kennedy had in the White House, and a staple among CV 67 commanding officers—the admiral started unspooling sea yarns. Marzano leaned in, amazed by the man's memory. Cherry joked that he had enough trouble recalling stories from last year, but here Buddy Yates, almost twice his age, retrieved memories from more than a half century before, in vivid detail. Marzano wanted to learn all he could about starting a new warship, about getting it ready under such scrutiny, consternation, and general uncertainty.

While the second *JFK* ship featured a whole bunch of new technology and gadgetry, Admiral Yates had carried something aboard his *JFK* that no other carrier at that time had—a computer. He found someone in the navy, scheduled to leave the service soon, who understood how to operate a computer, and worked out an assignment for the officer to the 67. The navy planned to use the computer to help the supply corps figure logistics for the *JFK* and its armada, as well as other carriers and their armadas. With the computer, they developed an information control center to manage the wares.

With that in mind, Yates told Marzano, "Get the crew on the ship and get them working. If they're on there working, they'll know enough when they're out to sea to know how to fix it—better than anyone else who could come along."

The admiral passed along another bit of wisdom: the name of the ship served as a great moral inducement for sailors to work hard. Only those in the navy or part of a navy family appreciated the importance and sense of tradition in the service of things like a ship's name. When you served on a ship, you became part of a new family with as much pride in that ship's name as in your own biological family's name. So, the association of CV 67 crew members immediately adopted Marzano and the soon-to-be CVN 79 plank owners into the carrier *Kennedy* family. The new motto became "Two ships, one *JFK* family." As part of tradition, Yates gave Marzano another replica of the rocking chair Kennedy used as a senator to help exercise his fragile back. The president took it with him when he went on Air Force One. The admiral also gave the future *Kennedy* CO a guide of sorts, a manual of lessons learned for bringing a carrier to life, his keys to success, which the first CV 67 CO had had bound. Marzano consulted that guide more than once in later months.

Following mentors took you only so far, however. Naval commanding officers, particularly those of carriers, needed to be more than carbon copies of previous bosses. A twenty-seven-year navy man, Cherry Marzano had transitioned from flying F/A-18s to the aviation nuclear officer program at his twenty-year point. He had commanded a transport dock vessel that transported marines—the amphibious ship USS *Arlington* (LPD 24), and he had been the XO of the *Lincoln* as well as the *Nimitz*.

Actually, his job as second in command of the carrier *Lincoln* likely best prepared him for his future responsibilities as commander of the *Kennedy*. On a carrier at sea, the CO pretty much became mother, father, judge, and jury—the officer whose word was law and whose power pretty much ranked with that of a god. When it came to what occurred on the ship, the CO nearly never checked with anyone before making most major decisions, which seldom underwent scrutiny later. However, on a carrier in a shipyard, especially on one still under construction, the CO shared that power. That officer served more like the naval mayor of this developing sea city, in much the same way as Butler operated like the shipbuilder-mayor of the city under construction. In that role, Marzano focused on building relationships and making connections with Butler, other shipbuilders, and all those on and off the waterfront who were turning these tons of steel into the nation's biggest warship. The soon-to-be *Kennedy* CO understood the shipyard and the organization—and how to build those relationships. In this job, that proved as important as developing that new positive command culture and climate.

Marzano focused on doing both. To inspire CVN 79 sailors toward the sense of service, commitment, cooperation, and courage he wanted, the new carrier captain found the perfect person to provide a personal and professional pattern: ship namesake John F. Kennedy. He created a PowerPoint presentation of Kennedy's life, naval service, and political career, focusing on the former president's sense of duty to serve the country and other human beings. Marzano used the presentation during Coffee with the Captain welcome-aboard meetings he had with very small groups of new sailors, offering coffee and chocolate-chip cookies baked by his wife while he took the time to talk to each of the crew in a relaxed setting and convey his command philosophy. The ship's training department started a "Camelot University" course to familiarize sailors with CVN 79's programs, with a segment dedicated to the life of President Kennedy.

Cherry Marzano sought another source of inspiration—the original carrier *John F. Kennedy* and those connected to the ship, recalling Buddy Yates's words. The name of the ship alone was a great moral inducement. Marzano cultivated a relationship with another CV 67 alum, Bob Haner, an original *Kennedy* sailor and the president of the CV 67 crew association. Haner became one of many key contributors who helped Marzano with one of the most daunting and stressful tasks to be tackled by the future CO: creating the official ship seal. The navy treated seals with reverence; each was a declaration of what the ship stood for and a talisman to protect the ship for decades to come. Marzano wanted something to tie CVN 79 to CV 67. He wanted something to capture the essence of Kennedy the naval officer, the president, the source of national inspiration. He needed it to pass muster not only with the navy, but also with the Kennedy clan, starting with Caroline Kennedy.

Cherry Marzano remained focused on the seal even after he officially became the first CO of the *Kennedy* on October 1, 2019. That very same day, the commanding officer and his initial crew of several dozen handpicked plank-owner sailors, officers, and chiefs moved into Building 608, right off West Street near Thirty-Fourth, across from the 7-Eleven, kitty-corner from Eddie's and just down the street from the Naval Sea Systems Command Newport News office and the city bus terminal. A grayish metal rectangle—long sides running perpendicular to the ground—Building 608 looked like any one of the scores of other similar structures on or near the waterfront, except to Marzano and everyone else "on board" CVN 79. Officially called the Pre-Commissioning Unit building, it essentially served as a land-based part of the *Kennedy*. Administratively operated like the carrier, this building would be the nerve center for the core crew as they developed their "esprit de carrier." Etched in the staircase between floors—in this building, they called them decks—were the ship's motto and the command's core values. The sailors and officers would work from there until they started to take over actual spaces in their ship.

On that morning, the forty-three plank owners lined up like a chorus, wearing their green cammies and tan ship baseball caps with a front view of the ship on the front, breaking through blue-and-white ocean waves, with USS *John F. Kennedy* emblazoned in gold lettering above the ship and CVN 79 below it. CVN 79 was also stitched in gold along the back adjustment strap. On the starboard side of the cap, the letters JFK stood out in red, white, and blue respectively. Above the letters, the cap featured the black silhouette of the original carrier USS *John F. Kennedy*'s mast with three aircraft zooming skyward—an E-2 Hawkeye, an F/A-18 Super Hornet, and an F-35 Lightning II. The crew of the previous *Kennedy* had originated the red, white, and blue JFK design on the side of the cap.

In front of Building 608, the CVN 79 crew watched as a *Kennedy* sailor raised the ship's ensign flag, the first *Kennedy* ship pennant raised since the navy decommissioned the first *JFK* decades before. Dressed the same as his crew, Marzano addressed his newly formed command for the first time, highlighting the historic significance of the moment and outlining the task ahead of them—delivering CVN 79 to the fleet. He settled into his new office on the upper deck, with bare walls and shelves filled with books like George Wilson's *Supercarrier: A Dramatic Inside Account of Life Aboard the World's Most Powerful Ship, the USS* John F. Kennedy—the first one—and other books and memorabilia concerning JFK.

His empty wooden desktop gleamed in the fluorescent light. Marzano again eyed the seal, which would adorn the ship for its half century of life, and the CO wanted the design to be just right. He wanted everything about his carrier to be just right. If they got it right from the very beginning, then there was a better chance it would stay right, would keep the right cultural course through its fifty years. The new CO set a goal for himself—to meet with every plank-owner sailor personally, face-to-face during his Coffee with the Captain meetings. That would mean a lot of sailors as the navy built up to the total deployment crew in the early 2020s of nearly three thousand. The navy handpicked each of the sailors, chiefs, and officers for their jobs on the *Kennedy*, to start it off right, and Cherry Marzano wanted each to know what a unique opportunity—and responsibility—he or she had. Each sailor "attended" Camelot University to discover just what it meant to sail on a ship with that namesake. Culled from around the country, much like the steel, valves, and other raw materials of the carrier itself, these men and women breathed the soul into the ship in much the same way as the sister and brother steelworkers had built the carrier's body. They did it as one, together.

That familial bond started to form almost immediately, as the first *Kennedy* plank owners started to tour the ship and its dry-dock cradle. Going around, in, and through the ship, they not only came away with an appreciation of their jobs, but also of the craftmanship of the steelworkers. Few of them had seen any ship, let alone an aircraft carrier, at such a stage, or watched the steelworker-shipbuilders practice their craft. Hailing from Cincinnati, Commander Mike Prudhomme, the first CVN 79 senior medical officer, who waited to take over the nearly finished hospital, made his rounds humbled and inspired by the massive size of the ship. They built a ship like this from scratch; he sought to do the same in starting a new floating medical facility for about three thousand sailors. Sure, it would be outside his usual comfort zone, working on an operational carrier, but few doctors ever had a chance to do something like this.

Another *Kennedy* plank owner blown away by the sheer size of the construction, Machinist Mate Nuclear First Class Aaron Zevenbergen, welcomed his assignment to his first precommissioning ship job. He and his team focused on one of the most important tasks in bringing life to CVN 79—developing the training, operations, and daily routine of the nuclear spaces, the power and propulsion center. Tight-lipped about the kind of work he did, the nuke machinist even avoided the usual joke cliché: "I'd tell you, but then I'd have to kill you." Zevenbergen and the other nukes became among the very first to take over their spaces and start working aboard the carrier. By the time the ship went operational, everything they did—firing up the reactor, monitoring its core, or even turning a valve—had to be pure muscle memory. Another plank owner, Nuclear Electrician's Mate Chief Petty Officer Kevin Stambaugh, from Lake Odessa, Michigan, reported from the carrier *Washington* with thirteen years' experience. He grew anxious, yet excited, to build the training program for his department.

Another reason also compelled Chief Stambaugh to get everything just right on CVN 79. John F. Kennedy was one of those naval historical names that represented a great legacy and so created a high bar that he wanted to reach and maintain.

While excited to start their *Kennedy* jobs, the plank owners also dwelled on the series of challenges that had continued to bedevil *Kennedy*'s predecessor, the *Ford*, which shuttled back and forth between Newport News and Norfolk when not being tested out in the Atlantic. The *Ford* was a constant reminder of a carrier-construction program gone off course, but it also was a real-life floating encyclopedia of lessons learned for the *Kennedy* plank owners. Few realized this more than new CVN 79 Ensign Cheyenne Scarbrough from San Francisco, who served on the precom unit for the *Ford* before being tapped for *Kennedy*. The *Kennedy* come-heres treated her a bit like a seasoned pro. One of her refrains became, "It's a constant push, and there are going to be great times and hard times. You just have to work through it." Tapping her experience and that of others from *Ford*, Scarbrough and the other *Kennedy* sailors navigated the minefields more rapidly.

More plank owners arrived every day. Originally from Jackson, Mississippi, Legalman Chief Rasha Shankle had shuffled in and out of different Norfolk units over the past nineteen years, like many US East Coast sailors. Like other CVN 79 sailors, she came to view the ship and the shipbuilders with awe. To be a part of that, to be one of the first on the new *Kennedy*, meant a great deal to her. Master-of-Arms First Class Chief Petty Officer Kristin Dennis saw it the same way. Creating the *Kennedy* ship spirit started with them, and she wanted to be able to say, "We built this foundation."

Talking with other plank owners in their new land-based CVN 79 headquarters in Building 608, Boatswain's Mate First Class David Kuefler summed it up: "I'll get to be present at the christening. The sailors in the future are going to be looking at that and saying, 'That was so long ago.' But I was there in the beginning. I'll be able to say to the kids, 'Yessir, I'm a plank owner.'"

Watching the *Kennedy* take shape in the capable hands of the shipbuilders, right in front of his eyes, simply knocked the socks off the St. Cloud Minnesotan. As he observed, "You usually only see things like that on the Discovery Channel."

Some things on the *Kennedy* were moved off the construction script in the eyes of the steelworkers still doing the hotwork and painting aboard the ship after it had survived Dorian's worst with little damage. Trades workers knew they had failed to complete the touted milestone work as indicated in public reports. They rushed through and closed tanks before thoroughly inspecting some of them, saying, "We'll patch it up and fix it later." To some, it seemed as though they built a Hollywood mockup of some spaces, like a fake Western town with phony storefronts. Painters left spaces with signs warning hotworkers to stand clear until the paint dried hours later. They returned and found the signs removed, way too early, with hotwork sparks flying within.

They removed all the ship weight possible—"every unneeded nut and bolt," steelworkers told one another at the union hall—to keep the ship from weighing too much. They made some alterations on the ship in the dry dock, just in case. They shut down ship bathrooms, fountains, and other water systems; the steelworkers used port-a-potties outside the vessel—two for women and eight for men. To prepare for the christening and launch, they closed some parts of the ship, making it a longer trek to get on and off the vessel. Steelworkers started to take restroom measures into their own hands, literally; trades workers would occasionally find filled bottles in empty spaces.

While steelworkers raced to seal up *Kennedy* in time for launch, other yard workers boarded the *Ford* as it went back into the Atlantic Ocean for another round of sea trials after a major pierside maintenance overhaul. This was four days before the flooding of Dry Dock 12, the first time the *Kennedy* hull would touch water. It was a time of carrier reckoning for the yard, the navy, and the workforce.

October 29, 2019, dawned gray, cool, and breezy on the waterfront, carrying with it a tinge of November. In Dry Dock 12, steelworkers readied and rechecked the ship for one of the most important milestones in *Kennedy*'s young life. While the Navy and Pentagon brass still looked forward a month and a half to the christening, this day caused much more stress for those on the waterfront. Today lacked the pomp and circumstance expected to mark the christening. Today, the yard planned to flood the dock and submerge the keel and about half of the bottom of the ship. Any serious problems today and forget about the christening and commissioning.

Butler, shipbuilders, and Marzano descended the blue steel construction stairwell into the dry dock through the early morning chill. The great ship rose before them, now looking like the fearsome carrier it was meant to be. The gray, black, and red hull looked battle-ready. The fairwater work was done, and the shafts and four propellers stuck out of the aft port and starboard sides, the part of the ship closest to land, like giant metal pinwheels. Weighing fifty-six thousand pounds (twenty-eight tons) each, made of bronze and encased in a shroud fitted around each of the five blades, the propellers rose taller than most of the houses in the area. Steelworkers removed the shrouds, which protected the propellers until they were submerged. As the steelworkers walked under the *Kennedy*, eyeballing everything, they moved like munchkins with white hard hats. Those on the waterfront became accustomed to walking beneath the carrier. For the unseasoned, though, looking at that ship and looking at the concrete-and-wood pedestals supporting it, many came away dizzy, awed, and humbled. The *Kennedy* now tipped the scales at seventy-seven thousand tons, about three-quarters of its prime fighting weight. The dock-flooding was running about three months ahead of schedule, but that didn't mean the CVN 79 team would be able to deliver the ship to the navy three months ahead of schedule—or any earlier at all.

Cherry Marzano joined those in the bottom of Dry Dock 12 inspecting the *Kennedy*. Wearing a hard hat, safety glasses, and a black windbreaker over his cammies, he chatted for a bit with Butler, who wore a blue windbreaker over his blue plaid shirt. The two men now worked as one *Kennedy* management team. Marzano had given his first personal command coin to the CVN 79 director. Steelworkers scurried about. Dockmaster John Anderson, a seasoned shipbuilder for a dozen years, scanned the ship and those working around it to ensure the safety of ship, dock, and people during the flood. Shipyard sounds—always ear-splitting even with ear protection—echoed and reverberated through the dock and around the ship hull. It sounded like being inside a giant industrial vacuum cleaner on the highest setting, punctuated by blaring warning beeps and the banging and clanging of metal striking metal.

Workers inside the ship prepared to test tanks, welds, and spots along the bottom of the hull, as well as electrical systems aboard the ship, once the yard started to flood the dry dock. All the digital engineering, preparation, and installation indeed made this job easier aboard *Kennedy*. Almost every mechanic and electrician now used a laptop, making it easier to energize so many electrical systems after the flood.

One of those doing the electrical work, Matt Phoebus, an X-31 Apprentice School electrician, usually spent most of his days running and pulling cable through the *Kennedy*, more firsthand experience than he'd dreamed of for this first year in the yard. When he decided to go to the Apprentice School, he figured on spending most of his time in a classroom or a lab, as well as on the ball field. He had a baseball scholarship to play shortstop for the school's very competitive collegiate-level team. But this kind of experience shone like gold for the teenager fresh out of Chancellor High School, a couple of hours away in Spotsylvania County, at the northern edge of the Tidewater Trail. He counted on getting a great education and learning a valuable skill, but not so quickly, especially right on the waterfront.

He'd wanted to design things when he was a boy of ten in semirural Spotsy, as folks there called it, and he figured to work his way through the yard after Apprentice School into marine design or marine engineering. Right now, he learned the ropes—well, the cables, and the power boxes, and all the ins and outs of shipbuilding trades interaction. Electricians worked with everybody—welders, painters, sheet metalists, pipefitters—all the shipbuilders. That kind of background promised to serve him well for what he planned to do later at Newport News Shipbuilding.

He quickly buddied up with the most seasoned electrician in his crew, a seventy-seven-year-old who knew all the tricks and who told him stories of the hazing newbies like Phoebus endured in the old days—like throwing hard hats into the James for the apprentices to retrieve—a fireable offense these days. One X-31er did tell Phoebus to get a "cable-stretcher" from the tool room. The apprentice refused to fall for that one, realizing no such thing existed. Other more gullible rookies, though, trekked off the ship to the tool room, only to return much later a bit embarrassed. The way the other electricians knew exactly where they were on the ship at any given time amazed Phoebus. For him, it sometimes seemed like one of those farmer cornfield mazes, only made of metal. He realized the carrier was big, but he'd never realized how big. And so many compartments. Compartment after compartment after compartment.

He loved it. He had the bug. He represented the next-generation shipbuilder, the paperless generation with a laptop and imagination. In the old days, as an apprentice, he would have been carrying around armloads of blueprints and spending most of his days shooting studs or grinding metal. But now, thanks to the digital waterfront, he spent his days wiring the ship. He certainly understood the concepts of modern digital shipbuilding. He grew up using something akin to a laptop in middle school. It made learning his new craft now so much easier, just as it made everybody's job this day much easier.

Leaving the dry dock by the metal staircase on the day of the flooding, steelworkers caught sight of the port side of the aircraft carrier, its propellers suspended above the concrete floor like mammoth fans, the hull stretching out toward the James and the flight deck jet elevators overhang blocking out the sky—the ship seemed impatient, as though it wanted to get into the sea now as it was built and meant to do. Perhaps the great war journalist Ernie Pyle put it best in a column he wrote in March 1945:

> An aircraft carrier is a noble thing. It lacks almost everything that seems to denote nobility, yet deep nobility is there. A carrier has no poise. It has no grace. It is top-heavy and lopsided. It has the lines of a well-fed cow. It doesn't cut through the water like a cruiser, knifing romantically along. It doesn't dance and cavort like a destroyer. It just plows.... Yet a carrier is a ferocious thing, and out of its heritage of action has grown its nobility.

Shipbuilders climbed out of the dry dock and walked down along the side of the dock toward the front near the river. They lined up along the side to watch, a line of hard hats leaning on the orange plastic barriers. Other workers lined the rails on the *Kennedy*, and still others stood behind the yellow steel railings on the outer edge of the caisson gate that separated the dock from the river. Below the white identifier DRY DOCK 12 painted large in white on rusty blue-gray steel and another layer of tan steel stretched a wide layer of brown-red steel—roughly the same color as the *Kennedy* hull—which featured twelve valve openings to the James, the size of massive manholes.

At 9:00 on the dot that morning, the yard alarm alerted everyone.

The valves opened, and torrents of water poured in, like small water-falls to start the ten-hour task of flooding the dock with about 160 million gallons, enough to fill about 242 Olympic-size swimming pools. Lee Murphy watched the tide of water spread from the river end of the dry dock toward the ship, with mixed feelings. This was the last time he would watch this as a shipbuilder; he planned to retire soon after the christening. On the side of the dock, white lines marked off the depth in feet. Workers stopped the flow at ten feet to allow the ship-builders to check the integrity of the hull, tanks, and other parts, in-spections they needed to repeat over the next couple of weeks, filling, stopping, checking, until the dock was filled with about twenty-seven feet of water. It took remarkably little time, about twenty minutes, for the first ripples of river water to reach the first bow pedestal of the ship. Only a few weeks remained before the launching. The carrier launch, timed and sequenced so carefully, with the James River water filling the dock bit by bit, little resembled the days when the ships slid down ways and splashed into the water.

Of course, the steelworkers at Local 8888 touted the accomplish-ment of launching the *Kennedy* three months early. But down at the union hall, President Charles Spivey focused on another milestone of sorts. As the yard hired those five thousand new workers over the past few years, Spivey, a former union organizer, sought to raise the local membership. In this year—the fortieth anniversary of the steelwork-ers' official entry into the yard and the twentieth anniversary of the game-changing 1999 strike—Local 8888 closed in on a membership of ten thousand. Spivey wanted to build on the success of recent con-tracts and force the company to provide better pensions when they renegotiated in 2021. Just a few short years ago, that 10K number seemed completely out of reach.

The yard dipped down to nineteen thousand workers starting in late 2015 into 2016 as the navy rescheduled overhaul work on the air-craft carrier *Washington* because of all the budget machinations going on in Washington, which delayed that carrier work for almost a year and pushed back the delivery date for the *Kennedy* to 2022. All that created a little valley for the yard's carrier workforce. Now, though, the yard anticipated a workforce closer to thirty thousand.

In the union local's newsletter, the *Voyager*, that fall of 2019, Spivey reminded workers that their recent 3 percent paycheck increases resulted from the 1999 negotiations.

Spivey put Kenneth "Chan" Lewis of 054/X-36 in charge of the organizing committee. A former 8888 vice president, Chan and other organizers hit the front gates in the morning, made phone calls in the afternoons, and knocked on doors on Fridays. Their message was simple: "It's wrong not to belong." The membership numbers climbed. The early launching of the *Kennedy* seemed to put the carrier program on a more solid course. More carriers meant more steelworkers. A blip, though, flashed on the radar, and again, the focus ping-ponged back to the *Ford* and those elevators. *Kennedy* work suffered, again.

11

BLAME GAME

FALL 2019

As the yard basked in its success with the carrier *Kennedy* in the fall of 2019, the *Ford*'s problems continued to drag down the whole waterfront carrier effort. A fix for the weapons elevators eluded engineers, and that rankled the navy brass, especially Secretary Spencer, whose "fire me" offer still hung in the air. The same week the yard flooded Dry Dock 12, Spencer flew over the Atlantic to board the *Ford* and light a fire under someone's butt to get those elevators fixed.

Spencer had told President Trump at the 2018 Army-Navy game in December that failure to get those elevators in order soon would be a fireable offence, and soon afterward, the yard had told the secretary to expect working next-generation weapons lifts in July 2019. Then, in the spring, yard officials told Spencer not to expect working elevators until late 2020 or early 2021. Spencer later called that a "moment of inflection." He called Huntington Ingalls Industries board chairman Tom Fargo. "Does the board of directors know what's going on with management here?" Spencer asked. "Because our trust and confidence in this project, the specific project of the elevators, has been eroded significantly." Spencer and Hondo Geurts took matters into their own hands. They issued twenty-two thousand work orders over the next three months to jump-start the elevator work. Seven elevators worked, and the navy certified four of those, but the work still lagged way behind schedule.

After being briefed on the *Ford* in the Atlantic, Spencer helicoptered back to the navy base in Norfolk to publicly vent to a circle of reporters. He looked like a shipbuilder himself, wearing a blue plaid shirt and a blue baseball cap with a gold-and-white N on the front and the word NAVY in gold beneath it. Despite his appearance, he gave shipbuilders no quarter, as he laid total blame for the elevator mess right at the feet of Newport News Shipbuilding management.

When the yard management told the navy in March that shipbuilders would need a lot more time to fix the elevators, Spencer sensed the yard remained unsure of the nature of the problem. If true, the potential existed for the same problem in *Kennedy* and every other Ford-class carrier. "That was a bit of a gut blow," Spencer said.

Except for its elevators, the *Ford* performed admirably during sea trials out in the Atlantic. It needed to prove its ability to reliably launch and recover aircraft, and the ship did just that for months. The real problem came down to publicity, not operations. Nothing sparked good press about a carrier more than an embark—a media trip out to the ship—to watch fighters take off and land. Embarks created headlines, pictures, and purple prose about roaring jets, not stuck elevators. Yet for some unexplained reason, the top navy brass and the Pentagon rejected such trips. The ship's successful sea trials went largely unreported, except by the trade press.

Spencer remained frustrated, impatient, and more than just a little ticked off. After all, he'd put his job on the line in promising working elevators, and he saw no indication of their appearance soon. Spencer's public bravado about the work matched the grandiose manner with which many Trump appointees conducted their jobs. But any previous navy secretary, or any other high service official, would have warned him in no uncertain terms: don't guarantee any kind of shipbuilding schedule, and certainly don't bet your job on it. Shipbuilders were not lying SOBs—far from it. Building ships was part art, part science, and no little bit of luck, good or bad. The best-laid plans often went awry, particularly when introducing different technology and dealing with aircraft carriers, the most complex weapons systems ever designed and built.

Spencer tried to rewrite the history of Newport News Shipbuilding's handling of the elevators. Think back to Manazir's tour of *Ford* earlier in the decade. The navy and the yard had been up-front with each other since the beginning about the risks involved in developing those advanced elevators, and they'd agreed to share the risks. Early on, the yard had wanted to develop prototypes to get a better handle on the elevator technology, but the navy rejected those requests in order to save money. On the other side of the coin, the yard had exuded confidence about building the new elevators; it had been building elevators on carriers for decades. Naval officials had shared that hubris. Both sides had failed to recognize the true complexity of the problem until much later, relying too much on their shared experience in building nuclear aircraft carriers, and also on digital engineering to compensate for major unknowns.

As Hondo Geurts told reporters in the Pentagon while Spencer publicly berated the shipyard managers, "Everybody underestimated the complexity of that, from a construction standpoint."

Knowing they were not the bad actors that Spencer had painted them as was cold comfort for the Newport News shipbuilders. It was never good to be a navy secretary's whipping boy when you made your living from the navy. What truly galled them, though, was the public way in which Spencer had blasted the yard. It just was not done that way, not for something like this, not for a construction problem that was tougher than anyone had thought it would be.

Then Trump decided to reverse a military justice ruling on a former SEAL named Eddie Gallagher, and everything changed for Secretary Spencer and the *Kennedy*.

In the spring of 2019, several fellow SEALs from Gallagher's own team reported to the navy that he had shot civilians and killed a captive Islamic State fighter with a hunting knife during a deployment in Iraq in 2017. They also said he threatened to kill the SEALs who had turned him in.

The court acquitted Gallagher of all charges except one—bringing discredit to the armed forces by posing for photos with the teenage captive's dead body—and the navy demoted him. Trump inserted himself on November 15, 2019, reversing the demotion. The navy also started disciplinary proceedings to strip Gallagher of his trident pin, the navy talisman marking him as a SEAL, and kicked him out of his unit. Trump tweeted: "the Navy will NOT be taking away Warfighter and Navy Seal Eddie Gallagher's Trident Pin. This case was handled very badly from the beginning. Get back to business!"

More than just about anything, the military brass hated when civilian leadership mucked around in the dispensing of military justice. Speaking with others in the Pentagon, Spencer threatened to quit, again, over the matter. With that in mind, Defense Secretary Mark Esper and General Mark Milley, the chairman of the Joint Chiefs of Staff, asked Trump to permit the disciplinary process to take its course. Apparently at the same time, Spencer used a back channel to tell Trump that if the president allowed the navy to go through the disciplinary process, the navy secretary guaranteed Gallagher's spot with the SEALs. Esper threw a fit when told about this, and he secured Spencer's resignation.

More than a few in and out of the Pentagon questioned whether it was really Esper's fury that brought on the resignation, given that many in the administration used back channels to Trump. Also, the administration had been putting pressure on Spencer to award a new frigate contract to a shipyard in Wisconsin for political reasons, but he said he'd rather be fired than make such a guarantee.

Spencer's resignation letter on November 24, the Sunday before Thanksgiving, put the blame not on Esper, but on Trump: "Unfortunately, it has become apparent that in this respect, I no longer share the same understanding with the Commander in Chief who appointed me."

For Gallagher, Spencer's resignation meant the former SEAL could definitely keep his trident. Esper felt it would be impossible for Gallagher to get an impartial hearing. For the yard, it meant a reprieve from Spencer's elevator wrath. Trump announced that he planned to nominate Kenneth Braithwaite, the ambassador to Norway, as the new navy secretary. Meanwhile, Navy Under Secretary Thomas Modly became acting navy secretary. During the recent public skewering of the yard by Spencer over the elevators, Modly defended the shipbuilders at a conference of military reporters and editors, saying the *Ford*'s problems started because of all the technology and new systems hastily squeezed into the ship.

Yard officials figured they'd found an ally, or at least a sympathetic ear. At the same time, Admiral Michael Gilday, the relatively new chief of naval operations, touted the need for naval aviation in the future in December 2019 at the Defense Forum Washington, hosted by the US Naval Institute. "We need an aviation platform," Admiral Gilday told the attendees. "I think there will be a requirement to deliver seaborne-launched vehicles through the air, delivering effects downrange."

Things were looking up again for carrier programs, but the Tidewater steelworkers still worried about the consequences of those unfixed *Ford* elevators. The yard also started to feel even more pressure from members of Congress to fix the *Ford*, but in a different way than with earlier problems involving ship systems, seen mostly as the navy's fault. On the *Ford*, that had been the case with the launch-and-recovery systems and the radar—and the decision to introduce all the new designs on one ship. The government created all those things, so most if not all of the problems associated with that equipment could not be traced back to the yard and steelworkers, who simply installed and integrated them on the ship. Through all the bad headlines that dogged those technologies, and the accompanying increase in sticker price, Newport News had kept its reputation intact. Folks in Congress and the Senate saw it as a navy problem. The navy failed to reckon on the true difficulty of the new technology and the costs to get it all right. But the elevators—the yard did that work. A failure there was a failure in shipbuilding, and lawmakers started to question the honesty and integrity of the entire *Ford* program, including *Kennedy* and later ships. Congresswoman Elaine Luria, a Democrat from nearby Virginia Beach, publicly called the *Ford* a "thirteen-billion-dollar nuclear-powered berthing barge."

To fix the *Ford*, the yard again cut into the *Kennedy*—at a time when the latter needed all the workers, equipment, and resources possible, to get it shipshape for its christening in a few short weeks. *Kennedy* machinists noticed the short-shrifting as they tried to finish their spaces. They ground and prepped an area to ready it for Jordan Patterson or some other painter to come in and coat it. They ground it to bare metal for final welding. But the yard had sent so many *Kennedy* workers to *Ford* that the bare metal rusted.

If shipbuilders couldn't finish spaces, they couldn't sell them— turn them over—to the navy. That transfer was always what sealed the bond between the steelworkers and sailors who now breathed life into the *Kennedy*. Shipbuilders enjoyed the pride of showing off their work, knowing they'd built everything right—every pipe, every cable, every paint stroke. Sailors loved taking over a space, calling it their own, like moving into your own first apartment.

Once sailors took over the space, it truly became navy property, no longer a part of the *Kennedy* still being built. It may sound a bit weird, but in the carrier-under-construction, you had parts of the ship belonging to the yard and other parts belonging to the navy. Shipbuilders no longer walked into a navy space to work. Even when they started navy-approved work in a certain space, they only did *exactly* what was on the work order. At the union hall, they joked that, in those spaces, "You have to get another piece of paper every time you pick up a wrench." And even if you presented the proper piece of paper, other issues arose. "It's like, 'This has to be approved by that person, but we can't find that person.'" If you needed to unscrew something to get to the work at hand, you needed another piece of paper. Find another something that's awry, and you grabbed both a yard engineer and a naval engineer to look at it. You often needed both engineers to approve even the tightening of a loose screw.

However, both the yard and the navy sought to turn over compartments, and plenty of new sailors kept streaming into Newport News to fill those spaces. When they got there, they were all "electronically greeted" by Information System Technician First Class Chandler Ragland, who created all the personnel records and accounts for the incoming *Kennedy* sailors. Every ship has a first sailor, and on *JFK*, that honor fell to Ragland, a Southern good ol' boy from Montgomery, Alabama. The twelve-year veteran, who'd joined the service at the ripe old age of nineteen right after 9/11, was proud of being the *Kennedy* number one. Yessir! He bragged about it to his nine-year-old daughter and anyone else who'd listen.

He brought his wife and daughter to Tidewater from Hawaii, where they had lived for six years, a typical navy family accommodating his sailor's life—a life he loved, although he missed watching Auburn games. Back home, as a small-town country boy, he had only dreamed of making a living by sailing a navy warship out of Pearl Harbor. He and his wife grew up in Alabama and she followed in his wake. He reckoned to make Tidewater his home for another six years as he focused on his naval career, getting those new *JFK* sailors squared away. He did more than create that electronic gateway; he also tried to set the right example and help develop that right CVN 79 culture.

That meant a lot to Damage Controlman First Class Caleb Peterson, a ten-year navy veteran who had never served with a crew from the keel up before. He stared at his tan plank-owner's cap and pictured himself on his deathbed, that cap hanging above his head, and saying, "Yeah, I was part of that." Like others, he walked beneath the ship and stared in disbelief at the size of the rudders and propellers. Because *Kennedy* sailors were still unable to work aboard the ship, he put his body and soul into setting up the right plan to protect the ship and keep it fighting even if damaged. He intended to do his level best to create an atmosphere that was different from that on any other ship in the navy.

Handpicked for the *Kennedy* posting, like the other plank owners, he had also picked the ship from his own set of choices, in no small part because of the name. The name meant something to America and, therefore, something special to him. He worked as hard as any farmhand back on his family farm in Durant, Oklahoma, to do his job just right.

Working as a kid on his granddaddy's ranch and dad's farm, he knew he wanted a job with a uniform—police, firefighter, EMT. He chose the military. He fell in love with the service and his job. He was off the farm and out of Oklahoma, heading to places like the Persian Gulf.

He took Oklahoma with him, of course. It came through in his broad, easy voice that could easily be heard a country mile away. He spent his first Christmas away from home sitting by himself on the beach in San Diego, in cowboy boots, swimming trunks, and a straw hat. He still loved to watch rodeo. Now and then he'd rope the ball hitch on the back of his truck. Sometimes, for laughs, he'd rope his daughter. He always kept a couple of ropes in the back of his truck. When not roping or watching rodeo, he liked to load up his boat and go fishing, sometimes for days. He could watch his line for hours and not move a lick. He fished locally at Rudee Inlet, down by Virginia Beach. He preferred to fish little creeks with fewer folks about. He still got a kick out of going with his daughter and son to Bass Pro, although the ones around Tidewater seemed so small compared to the ones he'd gone to as a kid back home—some big enough to hold two or three *JFKs*.

Only *JFK* mattered to him now: CVN 79. Years from now, long after he walked off the ship for the last time, no one on the ship would know who he was. But he hoped that, without even knowing it, they'd see something of him wherever they turned. He intended to leave behind something much bigger than himself.

He wondered if he would still be attached to the ship when it went to sea in 2024 or so. Like all the plank owners, he wanted to be aboard when the *Kennedy* hit the waves for the first time, but most realized how unlikely that was. While he loved his family and being near them every day while the ship was still stuck in the dock near their home, he was a sailor, and he wanted to be at sea with a deck rolling beneath him.

Lieutenant Commander Valerie Greenaway, the *Kennedy* training officer, understood the zero chance of her still being on the ship when it first went to sea. But Greenaway remained determined to fully develop the drill package for whomever that officer would be. It seemed a bit counterintuitive to develop a training regimen for a ship and crew not even out of dry dock yet, but that was what plank owners like Greenaway did.

As one of six kids growing up in Cleveland, she liked living in the city and never once considered joining the military. The idea of wearing a uniform turned her off. Her whole family—dad, uncles, brothers—all worked as police officers. She figured to make it in life with her mind, having been accepted to Hunter College in New York City for premed. Not being from a family of means, she decided to stay in Cleveland. She worked two jobs while going to Cleveland State University full time, catching naps in her car. She wanted a better life than that, though, and so walked into a navy recruiter's office. She started her navy career as a military police officer. She hated that but was moved to other jobs, and she figured to stay in long enough to raise money for college. By the time she joined the other *JFK* plank owners, she had served for two decades, a lifer.

In no hurry to return to sea, she enjoyed her landlocked job and a nice place in nearby Poquoson, just twenty minutes from Building 608. A mom with four kids, ranging in ages from two to sixteen, she liked being in their lives every day—as much as they loved it when she deployed. Her commute included no tunnels or bridges to contend with, but she still struggled with Southern drivers during "weather." Growing up in Cleveland, she was used to driving in snow and on ice. Here, in Tidewater, they just played bumper cars. Even driving in the rain could be interesting.

The *JFK* job jazzed her. She had been a training officer for the carrier *Ronald Reagan*, one of her two favorite presidents—the other being John F. Kennedy. Reagan reigned as the president of her childhood years of the eighties—years she loved. She recalled the fall of the Berlin Wall as one of the biggest events of her childhood. Reagan inspired her. Her affection for him rivaled that for her grandpa. Kennedy, on the other hand, she recognized as a navy advocate who did a lot for equal rights, something dear to her heart.

Both presidents skillfully projected themselves on the screen, but they were quite different, not only as men, but also as carriers. For one thing, on the *Kennedy*, Greenaway would have her own centralized training classroom space, unlike on the *Reagan*, where the staff spread her classrooms out in odd places and put her office right under the flight deck. At times, on *Reagan*, the noise made it impossible to think straight. Eager to get aboard *Kennedy*, she now pulled triple duty to get everything ready for the christening, taking on the work of operations officer and public affairs officer as well. After all, Caroline Kennedy planned to be there.

Captain Cherry Marzano's mind also focused on Caroline Kennedy—and the rest of the Kennedy clan and protectors. He felt oh so close to getting the ship seal just right. Still, he joked with his friends and even his wife: "I've been in the navy for twenty-seven years. I learned how to fly F/A-18s and drive ships, but I never received training on how to create a ship seal." He approached people he trusted and respected, inside and outside the navy, even his nieces, saying, "Give me some ideas." He received feedback from hundreds. He realized he'd nailed it when he got the thumbs-up from the Kennedy family. Yes, he also required navy approval, but he cared most about what the Kennedys thought.

Anchoring the seal, which would also be used for the official ship coin, was the same image of the Ford-class ships that dominated the front of the tan plank-owners' cap, the carrier plowing through the frothy blue-and-white ocean surf. Below that ship appeared the ship's motto, arguably the most important part of the seal: SERVE WITH COURAGE. Marzano came up with that motto after spending the summer researching and contemplating. For him, it summed up the man they honored, the mission the crew embraced, and the mantra they all lived by. Marzano explained in the official press release unveiling the seal on November 6, 2019: "From the first day of his presidency, he challenged every American during his inauguration speech to 'Ask not what your country can do for you; ask what you can do for your country.' He regarded serving one's nation as an honor and held the utmost respect for those who did so with courage, especially when faced with adversity. John F. Kennedy displayed extraordinary courage, both in combat as a naval officer, and as president of the United States."

Marzano had created a seal he felt honored John F. Kennedy, his service to the navy, and his vision for space exploration. A full-moon backdrop frames the carrier and a head-and-shoulder image of the former president. Above the moon appears the Roman numeral CIX, a tribute to Kennedy's heroic naval service as commander of Patrol Boat 109 in the South Pacific. Thirty-five stars form a half circle on the bottom half of the seal around the roiling surf, symbolizing the fact that JFK was the nation's thirty-fifth president. The thirty-fifth star is after his middle initial, and the two gold stars between CVN and 79 show that the ship is the second aircraft carrier to bear his name.

The seal unveiled, everyone connected to CVN 79 turned their attention to the christening. Many hoped President Trump would honor the back-channel requests to stay away from the obvious photo op on December 7, the date now set for the event, Pearl Harbor Day. What a great day to break a champagne bottle on a new ship bearing the name of a World War II hero who became a president and a martyr.

12

CAROLINE, AGAIN

WINTER 2019

As November rolled into December 2019, the Tidewater breezes turned colder and more contentious, sometimes kicking up white-caps on the James and sweeping pine needles along the sidewalks past the storefronts hawking industrial boots, or legal services for those injured in the yard. The yard rang with sounds of seagulls, sirens, and swarms of Canada geese stopping during their southern trek.

In the squat brick union hall just a few parking lots away from the yard gates, Local 8888 steelworkers enjoyed an early Christmas, basking in the afterglow of the recent Virginia elections. Not only did they now have a Democratic governor and lieutenant governor, but thirteen of the twenty-three candidates endorsed by the steelworkers had won their races. The Democrats now also made up the majority in Richmond. It appeared to be time to make some changes in the capitol—raise the state's minimum wage, abolish the right-to-work laws, and protect Virginia unions. As Local 8888 President Charles Spivey noted in a *Voyager* message to steelworkers, "We, as the middle class, must be heard, and the time is now."

Erica Brinson, *Voyager* editor, wrote, "We took the election by storm." She spoke the truth. Steelworkers knocked on more than three thousand doors, canvassing, working the phones, and handing out campaign literature throughout the region. They conducted voter-registration drives at the yard gates.

Not everyone working the waterfront celebrated the Virginia election results and the new Democratic leanings in the state. Over many a barbecue, some talked about how to protect their guns—as some folks believed Dems planned to confiscate them. Some vowed to guard their guns with, well, their guns. Others loaded up their cars with their prized possessions and drove out to the houses of relatives in other states to store the weapons until the political climate shifted.

The election over, Brinson turned her attention elsewhere. The colder air rippling the James reminded her and others on the waterfront of the coming winter and the harder times about to face those less fortunate. Newport News had always been a bit of a magnet for the homeless in a way surprising for a smaller city. Many who had served in the military, worked for the yard, or just found their way there fell on tough times, seeking nightly shelter under the 664 bypass or thereabouts. Brinson's opening note for the November newsletter reminded steelworkers, "This is a time where we really show each other how important it is to give. When I say give, I don't mean money. We are blessed to work in one of the strongest industries in the world. . . . So, if you can volunteer your time to help out your local food bank, read to the kindergarten students, donate your used clothes that're just sitting in your closet collecting dust, all of those acts of kindness are what makes this time of year so special."

The steelworkers' desire to serve their community showed in many ways, from corporate offices right down to the waterfront. CEO Mike Petters, for example, asked that his annual base salary be cut from $950,000 to $1, the difference going toward a new fund to offer educational aid to the children of company employees. His wife, Nancy, taught preschool. Petters's move became public only in the company's proxy statement filed with the Securities and Exchange Commission.

Brinson and other members of her church liked to channel their giving spirit through the Peninsula Rescue Mission at Thirty-Seventh and Huntington, one of the nicer cinder-block storefront buildings a few blocks away from the yard and the union hall. Inside, at dinner time, men lined up for a free meal—fried fish, chicken, or some other hearty fare. They said prayers and thank-yous. The served and the serving treated each other with courtesy and respect.

Respect—Local 8888 president Spivey demanded just that for his members. The union crept oh-so-close to hitting that ten-thousand-member mark, about a hundred shy, and hoped to hit the number by the end of the year, even without a contract to renegotiate. They still had two years left on the current deal.

While the holidayish temperatures dropped into the 50s, inside the yard, workers still felt the heat to get the *Kennedy* shipshape by the christening and to also keep the sailors happy about the progress on their carrier. To get ready for the christening, steelworkers buttoned up the ship so it looked acceptable from the bridgetop view, so to speak. That required more brute strength. But belowdecks, the *JFK* sailors wanted their spaces done to move in and take over as soon as possible, but they wanted to avoid what their predecessors aboard the *Ford* had suffered through at this stage—the "Plywood Palace," temporary wood-encased spaces for those sailors to train in. The cold, damp, and half-finished rooms were no proper home to start carrier training.

Chief Dominique Sherrod, a logistics specialist with seventeen years in the navy, traveled but a short distance from his childhood home to start work as a plank owner for the *Kennedy*. A native of nearby Portsmouth, he well understood Tidewater and the sailor's life. His father served as a career navy man but held back from pushing his son into the service. A navy recruiter simply caught Dominique on the right day, but he stayed in for the long haul. He loved the camaraderie and the security his job provided to take care of his family. He planned to use his logistics training to start a post-navy career.

For the moment, though, he concentrated on the current task of getting everything ready on the *Kennedy*, getting all the office supplies and so on. That proved more difficult than it sounded for a ship not yet commissioned, a precommission unit, or PCU. They lacked official navy "instruction" or standing for the ship, forcing him to work everything through the Supervisor of Shipbuilding, which owned the ship's "credit card." Before coming to the *Kennedy*, he'd worked similar jobs on other ships, including three previous carriers—the *Washington*, the *Truman*, and the *Enterprise*. He performed his job much differently when he was at sea: if sailors wanted or needed something not on the ship, too bad. And everyone wanted their available supplies right away. He became accustomed to moving fast. On the other hand, in port, if folks on the ship needed something, he needed to find it, whether on the ship or not. Going out for supplies presented no problem in Hampton Roads—his home turf.

As plank owners and steelworkers finished fixing up the *Kennedy* for the christening and launch, managers made some last-minute decisions on things like the weapons elevators. The yard created a separate elevator-work trade to form teams to work from carrier to carrier. The yard tried to have it both ways, too—trying to fix the *Ford* while playing catchup with the *Kennedy*, still moving machinists and other metalworkers from one elevator to another on both. It became quite an undertaking on CVN 79, with parts of the deck staying roped off as steelworkers finished up work on the other carrier. Butler decided to slow elevator work on the *Kennedy* until they figured it all out on the *Ford*.

The yard also needed more steelworkers than usual to go on the *Ford* sea trials to fix elevators and address other issues as fast as possible. No one could force shipbuilders to go to sea. Some workers could not; it would mean days on end away from home and family, and some simply could not afford to do that. Those and other workers remained on the *Kennedy*—where they were sorely needed to meet the christening date, launch, and pier shift. Indeed, the steelworkers at the union hall talked about the yard's new parking ideas for handling work traffic once the *Kennedy* moved. Some three thousand workers would make the pier shift with the *Kennedy*. The final outfitting of CVN 79 made it too heavy for Dry Dock 12, so the shipbuilders would float the carrier downriver a bit to Pier 3. To accommodate that move, the yard hoped to distribute the commuter traffic by offering more spaces to the steelworkers in satellite parking lots, including one at the old *Daily Press* building just up Warwick Boulevard from Hidens, and by setting up a shuttle services of buses to take some shipbuilders from the parking lot to the ship and back.

Management in the offices and corridors of Newport News Shipbuilding worried about the list of folks possibly attending the christening. Those concerns heated up e-mail exchanges and telephone calls to and from Washington, DC, and all the way up to Massachusetts. While the Kennedys would, of course, be well represented at the event, protocol also demanded that the sitting US president also be invited to speak and bask in the political glory. But some in the Kennedy camp wanted President Donald J. Trump to sit out this christening. Trump's policies, personality, and beliefs were antithetical to JFK's at a core level. However, few presidents would want to waste such potential political capital in a state like Virginia, which, although tilted Democrat, still had plenty of right-leaning voters in and around the yard and Tidewater.

Like the rest of the country, Hampton Roads had continued to polarize during the first couple of Trump years. In the coming months that situation would worsen, as it would throughout America, but at the time of the christening, in early December 2019, many in Tidewater saw the news coming out of Washington—the Mueller Report, the threat of impeachment, and the rest of it—as happenings in a distant country.

All that divisiveness dissolved, though, on the morning of December 7. The Trump people never communicated whether they were coming, which essentially meant the president intended to miss this one. If the president planned to come, the Secret Service would have been in touch, early and often. Security, already tight, would have turned suffocating.

As the morning sun started to glint off the James, gusts of wind scattered squalls of leaves down the streets with wind-tunnel force. Cars lined up on Warwick, near the U-turn at Sixty-Third Street, to gain access onto Shipyard Drive three hours before the christening. The traffic streamed down the hill and into the yard parking lot, where steelworkers and their families spilled out of their cars, dressed for winter, with white, steaming 7-Eleven coffee cups in their hands. On the small bluff overlooking Dry Dock 12, neighbors living near the yard started to assemble, along with steelworkers and other waterfront workers, to glimpse the *Kennedy* and Big Blue through the tree branches. Steelworkers had drained some of the water from the dock to set the ship back on its pedestals so as to more squarely secure it during the christening and keep it from moving. One designer, who'd started at the yard that summer after graduating from Rensselaer Polytechnic Institute, stared wide-eyed, not at the carrier, but at the thousands of folks pouring into yard. One woman, her gray hair carefully coiffed, cooed with excitement. "Caroline is going to be here today. Did you know that? *Caro*-line is going to be here today."

Buses ran past police cars up and down the streets parallel to the yard waterfront, carting sailors and steelworkers from Hidens, the old *Daily Press* parking lot, and other satellite parking spots. A lone sailor stood at attention, in his dress blues and white cover, in front of Building 608. Other sailors, similarly dressed, streamed down West Street toward a row of buses.

Inside the yard gates, the area around Dry Dock 12 buzzed with excitement and anticipation. After all, this was the last big, public event for yard steelworkers working on this ship. The next major one, the commissioning, likely would take place someplace else, after the ship had long been out to sea and was about to head to its new homeport.

Today's christening likely would be *the* big milestone in the yard for the ship. Along the dockside, near one of the massive legs of Big Blue, a whole series of large tents sprang up with vendors offering funnel cakes, corn dogs, and fries. Patriotic music blared as steelworkers and sailors jostled closer to the ship through the crowd and under arches of red-white-and-blue balloons. A small sea of white chairs faced the forward starboard section of the great new carrier, looking like rows of genuflecting sailors in dress whites. The yard had decked out the *John F. Kennedy* in its christening finest, with red-white-and-blue bunting around the edges and streamers of small flags along lines that ran from flight deck to dockside. In the front of the ship, just above the bulbous bow, hung a white plaque, large as two steelworkers, with the letters CVN, in blue, affixed at the top, and below that, in larger red lettering was the number 79, which took up about two thirds of the sign. Below the number, in blue, was the ship's name, JOHN F. KEN-NEDY, and below that, also in blue, in smaller lettering were the words CHRISTENING CEREMONY, then in still smaller blue lettering was the date, and then finally, still in blue, NEWPORT NEWS SHIPBUILD-ING. Bunting emanated from either side of the plaque, as though the *Kennedy* were plunging through a red, white, and blue wave. Below the plaque and bunting was the stage from which Caroline Kennedy would strike the champagne bottle to christen the ship.

Mike Butler relished all of it. Though in some respects he treated this like another workday, he also took time to savor and to reflect. With this ceremonial custom, they tapped into a tradition dating back to when the first boats put to sea. Of course, some folks did things differently, depending on the beliefs of the times. Take the Vikings. They preferred to tie people to the logs or rollers under the ships used to get the vessels to the water. The log covers doubled as human sacrifices to the gods, splashing the ships with blood as they entered the sea after what the Vikings called "roller-reddening." The christening tradition shifted from red blood to red wine to champagne, the French being the first to use the bubbly. The added pressure made the bottle easier to break. Tradition broke, or at least paused, when Newport News Shipbuilding launched the battleships *Kentucky* and *Kearsarge* on March 24, 1898, and a group of Kentuckians threw bottles of bourbon at the *Kentucky* hull.

On Memorial Day 1896, the yard launched two ships, the side-wheeler *Margaret* and the pilot boat *Sommers N. Smith*, the latter named to honor the yard's superintendent. *Sommers N. Smith* slid down the ways and promptly rolled over—turned turtle, as they say—and sank.

Nothing as embarrassing happened before or after, but Butler, Doverspike, and the others still kept their eyes peeled and their other senses heightened. Especially today—for a married woman was doing the honors and the general sailors' superstition held that the christening of a ship should not be performed by a married woman, as that would doom the ship to ill fate. Well—presidents' wives had been christening Newport News Shipbuilding ships for about a century. Besides, they claimed a waiver from superstition for Caroline, as she already had christened a carrier *Kennedy*, albeit as a young girl.

Anyone who had seen that would never forget it, and folks witnessed it around the world.

The navy christened and launched CV 67, the first carrier USS *John F. Kennedy*, in 1967, on May 27, three days before Memorial Day, following a keynote speech made by President Lyndon B. Johnson, who had not only been Kennedy's vice president, but also a former officer in the navy. At the time, the Kennedys and Johnson barely spoke to one another as they vied for control of the Democratic Party. Johnson predicted, "In the year 2000 and beyond, this majestic ship that we christen today may be sailing the oceans of the world."

Just about every Kennedy showed up, and the star—besides the ship, of course—was Caroline, just nine years old and the youngest ever to sponsor a ship built at the Newport News yard. All three major news networks sent news crews—and *Paris Match* magazine sent a correspondent. Taking a lefty's stance, Caroline took aim and smashed the bottle on the second swing, with Jackie Kennedy clasping her hands by her mouth and John Jr. giggling in glee. The trio seemed to shine, dressed in white. Some thirty-two thousand workers, sailors, and others attended and cheered. It was more than a shipyard event, a national event, or a Kennedy event. It became a Newport News event. The kids at Briarfield Elementary School built a large papier-mâché model of the carrier *Kennedy*, topped with miniature fighters. One of the students there, second-grader Karen Nicosia, wrote:

> *Caroline Kennedy will be here, I hope,*
> *To crack a bottle on a boat.*
> *I wish I could meet you on your trip.*
> *Maybe I'll see you at the ship.*

And now, on Pearl Harbor Day in 2019, thousands of folks again came out to see Caroline at the ship, breaking another bottle on another carrier *John F. Kennedy*. A row of photographers packed the bleachers right next to the bow of the ship, hoping to get that "money shot," the exact moment she smashed the bottle, to compare it with the same picture more than a half century ago. The others, who lined up at the gates, grabbed food at the tents, filled up the seats, and squeezed into the makeshift bleachers, only wanted to be part of the moment. Local 8888 president Spivey attended, of course. The Murphys and Elliotts also came, of course, but some workers, like Patterson and McCann, could not get tickets. Some union members stayed away on principle; they felt management only touted the work of its "employees" at such events, never acknowledging the steelworkers by name. Most attendees sat too far away from the ship to really catch the bottle-bashing moment, but they could see it on the temporary large, concert-style screen. Images of the keel being laid, the island being landed, and other milestones flashed by, with interviews by the shipbuilders about their work. A clip of the first carrier *Kennedy* christening showed young Caroline hitting the mark. The screen moved through scenes of JFK's life as a president, sailor, father, and American futurist, as well as parts of speeches and other comments by the man being honored that day with another ship being christened in his name.

Kennedy's iconic voice boomed over the yard loudspeakers, from a speech he gave at the Australian ambassador's dinner for the America's Cup crews in September 1962:

> I really don't know why it is that all of us are so committed to the sea, except I think it's because in addition to the fact that the sea changes, and the light changes and the ships change, it's because we all came from the sea. And it is an interesting biological fact that all of us have in our veins the exact same percentage of salt in our blood that exists in the ocean. And when we go back to the sea, whether it is to sail or to watch it, we are going back from whence we came. . . .

A bit before eleven in the morning, the sun loosening just a bit of winter's steely grip, the VIPs started arriving toward center stage near the bow of CVN 79. Petters glanced over at the *Kennedy*, confident it would serve as a top ship of the line for the next half century. Caroline strode to the stage in a full-length dark coat, dark gloves, and a special, long red-white-and-blue scarf that was provided by the shipyard for the occasion, with CVN 79 stitched at both ends. The scarves adorned most of the attendees, adding a more patriotic flair. Caroline's husband and her kids, JFK's grandchildren, made their entrance. The invocation called on those present to reflect on the "day when Americans came together."

Boykin played master of ceremonies. Her voice rang throughout the shipyard. She reminded everyone of what the steelworkers—she called them by their name—had endured since making that first steel cut about a decade before, working in pouring rain, unrelenting humidity, and bone-chilling cold. Her voice rose: "Our shipbuilders are the only people on the planet who can do this." She then suggested that the carrier *Kennedy* served as a portal to the future of shipbuilding, in the yard and the nation, as Newport News Shipbuilding used CVN 79 to move more squarely into the digital age—just as JFK had moved the nation into the space age. When, as president, Kennedy invoked the New Frontier, the shipyard president said, he was not referring just to a place, "but rather a way of thinking and acting to move our nation forward to a country of new discoveries of science and technology." And now, she said, his namesake ship would be serving as a proving ground for a new frontier of digital shipbuilding. With every new foot of digital cable in the ship, she said, "We build freedom."

The crowd listened courteously to the other speakers that morning, but they came to hear Caroline Kennedy, and most leaned in to better hear her soft-spoken words. She spoke from her heart:

"War taught him that courage is the most essential quality in public life. I hope his example will be a source of strength and inspiration for the men and women aboard the ship, and their families, for years to come." She closed with her father's quote, the words echoing through the almost-silent yard: "My fellow citizens of the world: ask not what America will do for you, but what together we can do for the freedom of man."

When Caroline finished, Boykin walked with her over to the bow edge of CVN 79, joined by Captain Marzano, and pointed out the sweet spots on the white-mesh-encased bottle and on the ship, below the plaque and bunting. The yard whistle blew loud and long. Smiling broadly and taking a firm grip on the bottle neck, the seasoned christener, president's daughter, and retired ambassador assumed a strong stance—batting righty this time—and took a strong swing. *Bonk*, but no smash. She immediately recocked and took another swing—this time smashing the bottle and spraying champagne across the bow and out into the bleachers. Several booms sounded and a shower of red-white-and-blue confetti filled the air over the crowd like a patriotic snowstorm. A diamond-shaped formation of F/A-18s—the kind of aircraft that would be deploying off the *John F. Kennedy*—soared and thundered overhead.

With the confetti still swirling about, Virginia senator Mark Warner took some extra time to talk to reporters. He lamented the partisan divisiveness gripping the nation. The country could use some JFK spirit, he said. "We could use a dose of that whole notion of, 'Ask not what our country can do for each of us' individually but what each of us can do for our nation. And now, more than ever, that spirit is what we need—and what plays out, day in and day out, in this shipyard, that sense of teamwork—we need to bring some of that to the halls of Washington. There is nothing in the future Americans can't overcome or can't accomplish, but we have to do this together."

Nine days after christening, with Dry Dock 12 sufficiently filled, six tugboats guided *Kennedy* out into the James and downriver about a mile to the shipyard's Pier 3 for more outfitting and testing preparation. The ship still lacked its own propulsion, its reactors still dark. Shipbuilders had finished *erecting* the ship. Now they needed to get everything hooked up and working inside the vessel. They had built the foundation, walls, and rooms of the house, and now it was time to test the plumbing and lay the carpet. That night, Butler, like so many others on the team, headed home. As he drove around the rear of the dock toward the gate, he looked back. The sight so shocked him that he stopped his truck and stared. Nothing remained there. No ship, no staging, no stream of shipbuilders coming on and off the fantail . . . nothing . . . just a big open view of the night sky. Asked about the moment later, he recalled, "I have never been more profoundly moved by anything than I was by that view of nothing. I have also never been more proud of the shipbuilders who had made that sight possible."

* * *

Many travelers regard December as one of the best months to visit Hainan Island, China's largest tropical island, roughly the size of Taiwan, which sits just south of the Qiongzhou Strait, which separates the Gulf of Tonkin from the South China Sea. Perched at China's southernmost point, Hainan features a tropical climate that attracts beachgoers from all over the country and Southeast Asia to the seaside luxury hotels of its southern city of Sanya. But Chinese president Xi made a trip to Sanya on December 17, shortly after the *Kennedy* flooding, to check on the progress of his own carrier at the naval base there. The base remained a bit of a mystery for many Western defense analysts, although it was clearly visible from the ritzy ocean waterfront lodgings, right across the road from a shopping center with restaurants and a Starbucks.

Xi went to the naval base to preside over the country's first domestically built aircraft carrier's commissioning, and the second carrier in its fleet, putting the Asian giant in the elite club of nations with more than one. The ship would help China flex its maritime might, and the pressure was mounting on the US Navy, and Newport News Shipbuilding, to get and keep their new carrier program on track. The navy had a long history of counting on its shipyards to win wars, hot and cold, and it needed its shipbuilders to do so again.

13

ON COURSE

WINTER 2020

Not yet certified for flight-deck airplane tailhook landings, the carrier *Ford* made its way through sea trials in the rough wintry Atlantic off the Virginia coast. To get from the Hampton Roads shore to the ship, shipbuilders strapped themselves into seats on a navy helicopter. Because they flew in winter—and a cold January 2020 day, at that—the passengers dressed appropriately, squeezing into the bright orange Gumby suit, as sailors call it—a full-length layer of rubber that covered every part of the body, from head to heel, stuffed into it like a human sausage and looking like a flaming tick. The navy required everybody to wear one on a flight like this, even Rear Admiral James Downey, the new carrier program executive officer, who assumed the post when Rear Admiral Antonio retired a few months earlier.

It took several minutes and, for most, a few helping hands to get into the suit. The suit made it impossible to walk or even sit, and getting strapped into the MH-60 Seahawk helo with that ridiculously onerous five-point seatbelt cut off circulation like modern-day torture. Nothing left to do but smile, give a thumbs-up, and find some way to rest your helmeted head so as to keep the suit from cutting off the oxygen at your throat. The pilots and crew estimated a trip time of about an hour, maybe more. Quite a commute in such a getup.

Downey endured the flight out to the *Ford* to witness some take-offs and landings—and to get a firsthand look at those troublesome weapons elevators. Getting the *Ford* squared away became an all-hands priority for the navy, from Acting Secretary Modly on down. The brass and lawmakers alike wanted to see the afterburners lit in Downey's behind and in the posteriors of everyone connected to this ship—to change the current course, at least in terms of publicity. So, the Seahawk flight also included some orange-sausaged journalists—reporters for trade magazines covering defense and the navy. One such carrier embark with all the pictures and purple prose of soaring jets trumped a hundred boring reports of poor elevator reliability.

As the helo approached the *Ford*, the view out of the aircraft's small oval window highlighted the new design, with the island placed so far aft. On the flight deck, flight crew members scrambled from plane to plane, ducking under wings and kicking the tires to ready the aircraft for testing and ship certification: F/A-18 Super Hornets, E/A-18G Growler electronic warfare aircraft (the fighter with different wing pods), E-2D Advanced Hawkeye command-and-control planes, and C-2A Greyhounds. The Hawkeye functioned as a flying flight tower with a big dish on top, while Greyhounds, mini–cargo planes, ferried people and equipment. The crew made particularly effective use of the new-design in-deck refueling stations, moving more like NASCAR pitstop crews than carrier caretakers.

During later flight operations, green-vested sailors directed an F/A-18F Super Hornet into place along the catapult rail in front of the jet-blast deflector. Leaning properly into the wind, *Ford* cre-ated forty-nine knots along the flight deck, and even Rear Admiral Downey braced himself with a stronger stance to keep his balance. Wingnut was right, during those earlier sea trials, to fear being swept overboard. Forget about the old Irish saying about wanting the wind at your back. For carrier launches, you wanted it head-on. Crouch-crawling back a couple of feet from the engine-revving fighter, one of the green-vested crew shot his hand forward toward the ship's bow like a football ref signaling a first down. The Super Hornet roared and zipped down the front of the flight deck—the first full-afterburner catapult shot off the ship of an aircraft in a "five-wet configuration"—maximum fuel load, including external tanks. No steam rose from the catapult rail. EMALS worked perfectly.

The test pilots loved it. Flying Hawkeyes and Greyhounds, Lieutenant Dan Marsik waited for the usual gradual buildup of speed and power as he raced down the catapult rail. Instead, he felt a crisp, instantaneous snap and release with much faster acceleration. For the bigger aircraft, the Hawkeyes and the Greyhounds, there was no longer that second rough "bump" at the end of the launch as on the Nimitz class. Approaching the *Ford* for a landing, he lined up for the right angle of attack, quickly moving hands and feet to compensate for the speed of the ship, the speed of his aircraft, the pitch and roll of the ship, and the yaw of his plane. The pilots experienced no difference in their landings on the *Ford*, compared to those on a Nimitz-class carrier—a good thing.

Sailors led Downey to the hangar and the weapons elevator maw. Having shed his Gumby suit and other flight-deck protection, he blended right in with the rest of the crew in his green cammies and tan ballcap. He climbed ladders and ducked through hatchways like a pro. Inside the elevator bay, the wireless, cableless electromagnetic motor tracks shone like stainless steel. Thanks to the efforts of two hundred Newport News shipbuilders aboard and working seven days a week, even holidays, the navy had certified all the upper-stage elevators, turning them over to the ship's sailors. The steelworkers prepared to turn over another three, and with seven of the eleven certified, the crew had weapon-magazine access forward and aft. Once they secured use of the lower-stage elevators, sailors would be able to load bombs from the bowels of the ship and move them all the way up to the flight deck.

The working AWEs, he contended in interviews with the press, operated correctly. "They're working toward design. The ship is learning how to operate them and maintain them."

Downey realized that shipbuilders could have fixed the elevators faster if the navy had kept the *Ford* in port. But the service also needed to test its propulsion plant and the upgrades on other systems, as well as to conduct its flight certification. Pragmatically, it made sense to do the work at sea. As the carrier barreled its way through the waves, the hull twisted and torqued. The worse the sea state, the more the great, long ship bent. And the elevator tracks shifted as well. The best way to tweak the lifts for such movement remained to be found at sea.

Steelworkers now needed to complete those spaces to make sure they met the watertightness requirements for door and hatches. Once again, the yard needed more manpower, and once again, that manpower came from CVN 79. Lucas Hicks sat down with Mike Butler, who understandably saw things through a parochial lens when it came to getting work done on the *Kennedy*. Lucas told him, "Seventy-nine is never going to finish if seventy-eight doesn't first." If that meant putting a hundred additional people on *Ford* from next week to muscle through it, so be it. In the end, that only helped *Kennedy*. "It's a self-serving thing."

Plus—Hicks and the shipyard had something to prove. Even his bones told him CVN 78 was a good ship. The Ford class was the kind of carrier America needed to fight China, Russia, and whomever else dared to square off with the US over the next fifty years—and beyond. The Ford-class carrier was the most formidable weapon, the most complex piece of machinery, ever launched from the yard and put to sea, and he had no problem squeezing his big frame into one of those orange Gumby suits to take the press out to the *Ford* and preach the gospel.

As if to underscore Hicks's concerns, the Pentagon released an internal report shortly before Downey flew out to the *Ford*, noting that CVN 78's issues threatened to blow up CVN 79's costs and schedule. Back in the shipyard, on CVN 79 at Pier 3, Butler and his team turned over more spaces to the navy as the *Kennedy* crew started to take ownership of their boat, moving in, compartment by compartment. They began with the heartbeat of the ship, in its reactor department, as the nuclear sailors had begun to train and work aboard the carrier about two months after the yard launched ship three months ahead of schedule. Marzano liked what he saw. Recalling Buddy Yates's advice, Cherry saw the importance of getting *Kennedy* sailors working aboard their ship as soon as possible. Boarding the ship, manning their own official navy compartments, and doing their jobs—that helped the Reactor Department to settle into the day-to-day routine and started the *Kennedy* on course to becoming a navy carrier.

The naval officer directly responsible for those nukes was Reactor Officer Captain Joe Klopfer, whose ruler-straight frame, close-cropped hair, and American-pie countenance made him a recruitment poster come to life. Captain Klopfer's nuke crew prepared to take control of ten primary reactor systems, one every few weeks. He understood all he needed to do to bring a Ford-class reactor plant to life, having served as the main propulsion assistant aboard *Ford* at the ship's Reactor Department startup. He kept a headful of lessons learned—including, unfortunately, his own personal experience in the "plywood palace."

Like many early *Ford* sailors, Klopfer moved into the palace and suffered with everyone else in those belowdecks storage spaces, shivering in the damp cold and choking on the foul, poorly ventilated air—or whatever was in it—that condensed on the bulkheads. Everyone got sick. Once he got the top nuke job on *Kennedy*, he and other navy carrier officials approached Jeff Graveley, the yard superintendent in charge of spaces, and told him, "What we really want is our own space."

Gravely replied simply, "OK, we'll make it happen."

And that's exactly what he did on the *Kennedy*, providing the nukes with their own official Reactor Department spaces, months ahead of schedule. That not only made it easier and more efficient for Klopfer to train his sailors on the actual operational ship systems, but it also strengthened the sense of partnership between the yard and those who called the *Kennedy* their ship, called it home. The reactor captain wanted to avoid getting in the face of the yard workers; he wanted them all to face their challenges together. Even the smallest of details made a difference. For example, initially the yard planned to give the Reactor Department office the usual tannish carrier office-space decking; it was a struggle to keep it clean, or at least make it look clean. Klopfer asked for new flooring, a blue and off-white mosaic that wouldn't need to be swept or mopped every few minutes to make it look nice. Again, Gravely complied and that simple little change saved sailors work daily, if not hourly.

A special relationship develops between sailors and yard workers on a nuclear-powered ship. Unlike fossil fuel–powered navy ships, on which shipbuilders test and operate the equipment and systems, only specially trained sailors and officers can operate, test, and control carriers and their nuclear systems. Navy rules forbade shipyard workers from entering specific spaces after the turnover without proper authorization, even to turn a valve or check a breaker. Getting the nukes working on board as quickly as possible became a priority—nukes like machinist mate first class petty officers Brandon Nowlan and Evan Raycraft.

Born and raised in a small town in upstate New York, the kind of place that boasts of having three stoplights, Nowlan opted now to live in the small Tidewater city of Suffolk, just over the bridge or through the tunnels, depending which way you went. Either way, come summer or the final yard whistle, he faced traffic jams unlike any he had ever experienced before. And when it snowed, forget it. Four inches of snow would shut down everything. He braved the weather, though. He climbed into his Jetta and drove; where he came from, you didn't need four-wheel drive for four inches.

His family back home opted for either law enforcement or the military, and his elders compelled him to continue along those lines. He joined the navy in October 2009, hoping to be a search-and-rescue swimmer, but he failed the vision requirements. His ability to process information like a computer, though, prompted his naval handlers to push him toward reactor work. He connected with the Rickover ideals of dedication, leadership, and responsibility—holding yourself and others accountable for their tasks.

Another *Ford* transferee, he brought with him the wisdom of that ship's chiefs and master chiefs—the middle-management between officers and sailors—who him taught how to do things the right way. Now, having reached first-class petty officer status, he wanted to do the same on the *Kennedy*. Starting from scratch, as a plank owner, invigorated him—taking junior sailors and training them on nuclear basics to where he could see the lights click on in their minds. With the sailors now on a ship in the water, morale remained high.

Evan Raycraft grew up in Decatur, Illinois, and bragged that his hometown was the original home of the Chicago Bears, known previously as the Decatur Staleys. Raised in a strict household, he intended to be a surgeon like his dad, but college, it turned out, was not for him. When he returned home, his father gave him a choice—return to college or enlist. He joined the navy and never looked back.

Nuclear training sounded challenging and paid well, so he found himself training on the naval plants in Charleston, South Carolina, before heading out to be qualified on the carrier *Bush* in 2012, while the shipyard finished building the ship. He deployed on the carrier two years later and loved the sea life. He slipped easily into a daily rhythm in the Reactor Department, standing watches, training, and sleeping. At sea, you controlled your schedule and, for the most part, your tasks. Back in the yard, they threw just about everything at him.

Whether at sea or in the yard, he focused on performing *perfect* nuclear operations. He recalled one of his favorite Rickover-isms: You can't have perfection, but you can strive to have it. Certainly, you have integrity—something Raycraft tried to impart to his sailors. If something happened, if they made a mistake, he demanded to know about it. Don't try to cover it up, he told them. If you try to cover it up, there will be worse consequences. If I tell you to do something, and you don't know how to do it, ask. Don't spend hours wasting time, trying to look it up yourself. Someone can show you what you need to do.

What he did now on *Kennedy*, he believed, not only stuck with his trainees, but also helped build the lifelong culture aboard the ship. Negative attitudes now would ruin it for decades to come. They possessed the unique opportunity to carry on the name of the only president to receive the Purple Heart. Although, to be honest, Raycraft reminded folks, the PT captain did sink the first boat he ever commanded. Still, look at all that American history associated with the Kennedy name.

Nowlan felt the same. A navy namesake, Kennedy also was an American namesake. People just about everywhere recognized the Kennedy name. This ship would not only show American presence around the world, but it would also carry on tradition.

Kennedy's push for greater civil rights while he was president remains one of his legacies. On January 20, 2020, Martin Luther King Jr. Day, *Kennedy* Command Master Chief Thaddeus Wright read aloud to CVN 79 sailors President Kennedy's televised address to the nation on civil rights of June 11, 1963:

This Nation was founded by men of many nations and backgrounds. It was founded on the principle that all men are created equal, and that the rights of every man are diminished when the rights of one man are threatened.

That line resonated with everyone in uniform then, and always, and for Wright it underscored one of the principles of American democratic faith, a faith being tested just as much in 2020 as it was in 1963.

Today we are committed to a worldwide struggle to promote and protect the rights of all who wish to be free. And when Americans are sent to Viet-Nam or West Berlin, we do not ask for whites only.

As the command master chief, Wright, the senior enlisted sailor on the ship, set the tone for the development of the entire crew that day. His family experienced firsthand the battle for civil rights.

Wright, a true old salt, served in the navy for more than three decades before taking on the *Kennedy* post. As a command master chief, he became the sea-dad of sea-dads on the ship, the one to whom everyone turned for sage advice and model behavior. The *Kennedy* assignment was his third time working a ship in Newport News. He had been on the *Nimitz* in 1997 when the carrier came in for its refueling, and then he had come back as a master chief on the *Enterprise* about a decade later.

Kennedy—that name anchored a special spot in Wright's heart. Though he grew up in Elmira, New York, his family hailed from the segregated city of Greenville, South Carolina. They were among the Greenville Eight, the high school and college students who successfully protested the segregated library system in Greenville, South Carolina, in 1960—including the college freshman Jesse Jackson. Google the Greenville Eight, and you will find a picture of the eight, including Jackson, and there, next to the future national figure, stands Wright's father, Willie Joe Wright, and nearby, his mother, Hattie Smith Wright. His father had studied under Dr. Martin Luther King Jr. in college and went on to be not only a minister, but, like his mother, a teacher. As a toddler, the future *Kennedy* master chief saw pictures around his house of John F. Kennedy, Robert F. Kennedy, and Dr. King—several, in fact, of Dr. King and his father. He grew up assuming Dr. King was part of the family.

In a very real sense, he became a product of his parents and their generation; he achieved his position because he stood on their shoulders. When the navy named CVN 81 *Doris Miller*, he swelled with pride. He, too, started in the navy as a cook—stoking a yearning to learn the culinary arts—and he discovered that cooks found out more about the officers and crew than anyone else on a ship. A carrier now carried the name *Doris Miller*, and he had secured the job of the command master chief on the second ship called the *John F. Kennedy*. He saw symmetry, some kind of star alignment, in all that.

On Martin Luther King Jr. Day, Wright reflected on Kennedy's speech of June 1963. Since he had taken over as the CVN 79 command master chief, he'd found himself reflecting quite often on quotes from the ship's namesake. One that had hit him was from Kennedy's Thanksgiving proclamation of 1963. "As we express our gratitude, we must never forget that the highest appreciation is not to utter words but to live by them."

It's an attitude he lived by and wanted to instill—just as he believed Captain Marzano did—on the *Kennedy*. They wanted to do more than just utter core values. They wanted the sailors to live by them. In a sense, they sought to re-create the carrier in President Kennedy's own image. Wright and Marzano wanted to imbue the crew with a sense of service. The command master chief was certain his crew was more than up to the task. He saw it in their faces; they were not going to fail.

About a week later, on January 31, 2020, about a half hour away from the *Kennedy*, thirty-nine sailors from the Reactor Department lined up in the crisp morning chill on the wooden deck of the World War II battleship-turned-museum USS *Wisconsin* in Norfolk. Some wore dark waist-length jackets, but many wore their short-sleeve, button-down tan work shirts, despite the cold, with dark slacks and head covers. They faced the bow of the ship, in front of the ship's massive guns, facing out to sea, with family members and naval mentors standing along the port and starboard sides, and commemorated their reenlistment by again repeating their oath to defend the Constitution, their breath mixing with the Tidewater air. Back in Newport News, the shipyard prepared for another hiring spree, looking to add thousands to keep up with carrier and sub work. Little did they know how abruptly those plans would come to a screeching halt, upended by an outside force no one had foreseen or predicted—the biggest challenge the waterfront would face in building the *Kennedy* or any carrier.

14

COVID CARRIER

SPRING 2020

Normally, shipbuilder execs use the annual WEST 2020 national naval conference in San Diego as a prime opportunity to enjoy some camaraderie in the sun-splashed Southern California early spring while others in the colder mid-Atlantic, Northeast, and Midwest climes shivered in the March cold. Seasoned waterfront managers, retired navy brass, and shiploads of suppliers gather to grab drinks, trade yarns, and hawk wares. However, a sober and somber tone darkened the conversations that third month of 2020. While the Trump administration remained more interested in seeking revenge on Lieutenant Colonel Alexander S. Vindman for his impeachment testimony than talking about some odd new virus called Covid-19 spreading elsewhere, managers in yards throughout the US found the bug worthy of their concern.

From what they saw elsewhere around the globe, that bug gave them reason to worry. Indeed, American-based shipyards owned by foreign parent companies received red alerts in the beginning of the year about the potential production catastrophe awaiting the US shipbuilders. After the convention, Italian-owned Fincantieri Marinette Marine in Wisconsin stopped travel among its three shipyards and stopped all travel out of Wisconsin. Based on warnings from across the Atlantic from its UK-headquartered parent, Virginia-based BAE Systems Norfolk Ship Repair yard started to stock up on hand sanitizer and similar supplies.

By mid-March, it had become clear throughout the US that Covid-19 would disrupt society, the economy, and daily life. Throughout the nation, states shut down all but essential businesses, as determined by the federal government, including military production at the shipyards and associated suppliers. As Hondo Geurts told shipbuilders, the nation's defense took no breaks.

He spoke the truth, in theory. In 1918, during the Spanish flu pandemic, the Newport News yard shut down during the summer when temperatures reached 110 in the shade. Though the flu infected thousands at Newport News Shipbuilding alone, the government deemed the yard an essential business, and the waterfront continued to build vessels needed for World War I, even amid reports that area residents buried their dead in their own backyards. A copy of a memo from that time, dated September 19, 1918, wound up being routed later to Jennifer Boykin. It was from yard president Homer Ferguson warning shipbuilders about the dangers of the flu. It included advice from the company's chief surgeon to "avoid as far as possible, all public meetings or gatherings and other close contacts with persons sneezing or coughing." At the time, some called the Spanish flu the Hun plague, believing it to be a German plot to win the war.

Many on the waterfront found it hard to understand how worry about a cough or sneeze could wreak as much havoc, if not more, than a hurricane. But this virus threatened more damage than Isabelle or Dorian. Of course, those storms devastated, but shipyard managers understood how to prepare for them, how to survive them, and how to pick up the pieces and move on. Hurricanes damaged mostly property, and as costly as they might be, piers, cranes, and docks could be repaired or rebuilt. This new virus, however, ravaged people. It threatened to devastate the workforce, a far more difficult asset to rebuild.

Indeed, depending on how many became infected and how severe the infections turned out to be, it could set shipbuilding back decades. The uncertainty of it all wracked nerves. What should the yards do and when should they act? Trump claimed everything was under control going into March, just as state governments shut down everything. In Virginia, Governor Ralph Northam closed schools, demanded that folks wear masks, and urged everyone to socially distance, as requested by the Centers for Disease Control (CDC). In the Newport News Shipbuilding yard, something needed to be done, and right away.

Worrying about his sailors who had already begun to move aboard the *Kennedy*, Captain Cherry Marzano got a stark glimpse into how an outbreak raced through an aircraft carrier—from the other side of the world.

An American aircraft carrier pulling into a Vietnamese port was a rare sight. The mid-March arrival of the *Roosevelt* was only the second carrier visit to Vietnam since what Americans call the Vietnam War. *Roosevelt* steamed into the port of Da Nang just in time to commemorate the twenty-fifth anniversary of the start of bilateral relations between the two countries. At the helm stood Captain Brett "Chopper" Crozier, a seasoned carrier handler.

About two weeks after that port call, three *Roosevelt* crew members tested positive for Covid-19 on March 24 as the ship headed to scheduled port call in Guam. A helicopter flew the infected persons to a hospital near the port. The virus quickly spread, and Crozier, concerned that the navy was dragging its administrative feet in working out a plan to care for his crew—the sick ones and the ones he hoped to keep from getting sick—sent e-mails outside his chain of command to get things moving. In one letter attached to a message addressed to numerous recipients, he wrote, "We are not at war. Sailors do not need to die. If we do not act now, we are failing to properly take care of our most trusted asset—our Sailors."

His gambit paid off more than he had planned. Leaked to the press, the e-mail set off a national firestorm that upended the navy's highest echelons. On April 2, Crozier was fired, a decision later blamed on navy leadership's desire to head off Trump interference in the matter—the president expressed his support of the move. The navy relieved Crozier of command, officially, for going outside the chain of command, something Acting Navy Secretary Modly made clear when he flew from the Pentagon to the ship in Guam on April 6 to belittle and berate the commanding officer in front of thousands of crew members, who appeared on YouTube cheering and chanting "Captain Crozier" as the CO left the ship the day before.

Modly's voice boomed over the ship's intercom system and was taped by angry sailors who later released it to media outlets. "If he didn't think that information was going to get out into the public in this information age that we live in," Modly said, "then he was . . . too naive or too stupid to be the commanding officer of a ship like this."

Modly added that the alternative is that Crozier did it on purpose, saying that would be a violation of the Uniform Code of Military Justice. Sailors cursed loudly as the acting secretary ranted for fifteen minutes, portraying their captain as a man too blinded by his "love" for his crew to keep the overall safety of the crew, the ship, and the mission in sight.

Many in Congress and the Senate condemned Modly for his speech, and he resigned the day after he made it. Six days after that, a *Roosevelt* sailor died from complications due to Covid-19—the first active military fatality attributed to the virus—as hundreds of sailors tested positive and several required hospitalization.

Later that month, after the US Navy leadership had gone through a public meltdown, in another part of the Western Pacific, China's carrier *Liaoning* and its strike group steamed through the Miyako Strait and past Taiwan, cutting a course in the 155-mile-wide waterway between the Japanese islands of Okinawa and Miyako, then turning south and passing east of Taiwan. China was obviously trying to flex that carrier muscle, to close the gap between the US flattop forces and those now growing within the Chinese fleet.

In mid-April, the navy and marine corps and the Public Health Center and the CDC started a Covid-19 public health outbreak investigation, something that anyone who had any business dealing with carriers, even building them, would want to follow closely. More than 1,500 *Roosevelt* sailors would wind up infected. It became clear that the virus could spread like wildfire through a carrier-like community. The navy and yard needed to determine if that could be the case not only on operational ships, but on other carriers moored pierside, like the *Kennedy*, and assess the vulnerability of CVN 79 sailors and steelworkers.

Cherry Marzano decided to take precautions with his crew before seeing the results of any CDC carrier-Covid-19 investigation. Sure, he preferred to train his sailors and officers aboard the *Kennedy* to help them become one with the ship and infuse their spirit into the ship. But he opted not to risk crew members' lives for such training on board the ship under threat from such a dangerous and unpredictable virus. He ordered those on the ship back to Building 608. The *Kennedy* plank owners packed up what they needed from their compartments, closed up their spaces, and disembarked, grudgingly trudging back into their land-based abode. They would retake the ship when the time was right. Back on land, Marzano spread out his crew and created bubbles within bubbles of protection for those under his command. He erred on the side of caution, but he could afford to do so at this stage of construction.

Covid-19 dominated the minds of waterfront steelworkers on the James. After the Virginia governor closed the schools in March for the academic year, homes became the community classrooms, daycare centers and, with everything shut down, round-the-clock self-contained worlds. In Newport News, parents became frantic about childcare, especially those who worked in the shipyard, whose work the government deemed essential. "The conditions seem to change every day, along with the expectations placed upon us at work and the options we have to protect ourselves, our co-workers and our families," Local 8888 said in its April edition of the steelworkers *Voyager* newsletter. "Many of us do not know what to expect going forward." The union noted, "The shipyard will not shutdown. . . . We at the Union represent the workers to remind them that those things depend on our continued health and well-being as well as the risk we incur every day we come to work in spite of the CDC and Governors Guidance to 'Stay at Home.' . . . In order to protect ourselves and our families, particularly those at increased risk, many of us have reluctantly had to stay at home."

Yes, other businesses deemed essential voiced the same kinds of concerns, but none of them faced the scope of risk as those on the waterfront, with tens of thousands of workers, sailors, and others all cramming into one waterfront space. The yard offered some help straightaway, and the workers welcomed it. Writing in the newsletter directly to the union members, President Spivey pointed out, "A lot of you have taken advantage of the liberal leave offered by the company."

Liberal leave came with more than a few caveats. In essence, the yard offered workers vacation, personal time, or unpaid time to stay at home because of virus-related concerns. Some workers endured underlying conditions or came into close contact with people who did, but the biggest issue at most of the yards, executives said, concerned childcare. Schools, after-school programs, and other childcare centers closed throughout the country. When Virginia suddenly closed schools, yard workers from the waterfront to the boardroom pulled double and triple duty as remote teachers and homecare providers, with no alternatives. The yard gave people the flexibility to stay home, without suffering any consequences, providing they proved extenuating circumstances. The yard guaranteed their jobs, if the workers stayed out to take care of things.

"Many of you are exhausting your leave time to do this," Spivey continued. "We are not in a labor dispute with the company. We are in a national pandemic. Something we have never had to deal with before. Something unfamiliar to all involved."

Spivey voiced his concerns during local television news interviews. After some media misquotes, the international muzzled him. He faced criticism from all sides for voicing his concerns, but that came with the job. He believed they would all survive this and emerge better people working for a better company in a better union in a better country—"by the grace of God," as he put it in a *Voyager* message.

God very much occupied the mind of Mike Butler. He believed times like this made people face their eternity. For the adult Sunday School he had been leading for about a decade at Southside Baptist in Suffolk, they were studying, appropriately enough, the book of Revelation.

The most anticipated revelation in Hampton Roads, though, was what Newport News Shipbuilding president Jennifer Boykin planned to do at the shipyard to keep operations running while ensuring a safe workforce. Unlike Cherry Marzano, she could not abandon the *Kennedy* or any of the ships in the yard. She needed to act quickly but not rashly. She opted to rely on the collective wisdom of the smart, dedicated shipbuilders to chart a good, solid, safe course.

One of the first decisions she made was to be honest, open, and transparent with workers and the community about cases in the yard, infection concerns, and so on. She would be publicly berated several times for that decision, but the e-mails she received told her she did the right thing.

For work on the *Kennedy* and other ships, managers assessed work they needed to do, the workforce on hand, and what trade-offs they needed to make. They faced not only a shortage of labor, but also of other resources. In some ways, Covid-19 only exacerbated the focus on *Ford*, at the expense, once again, of *Kennedy*.

Before the yard took any measures, Boykin wanted to know what they needed to do to protect workers but keep navy work moving ahead and ensure the yard's viability when it all ended. She prepared for the worst and developed a plan to survive with the workforce and business intact. To get that stark true picture, she needed honest input from her seasoned pros. No other shipyard would be as taxed during this Covid catastrophe as the one in Newport News. While every yard braced for challenges, Newport News Shipbuilding faced a particularly complex set of concerns when Covid-19 outbreaks started. Steelworkers were still building *Kennedy*, refueling *George Washington*, and constructing attack and ballistic-missile submarines. While not in Newport News proper, the *Ford* still commanded quite a bit of steelworker attention, too, robbing critical workers and resources from *Kennedy*.

First, the yard made sure it identified its priorities correctly—and that those priorities meshed with the navy's. The yard focused its workers and resources on ships being fixed or overhauled—like the *Ford* and *Washington*, those attack subs being readied for commissioning, and ballistic-missile subs. *Kennedy* work ranked after all that. Those matched the navy's priorities.

For Boykin and others running the shipyard, the number-one concern remained ensuring workers' health and safety. Trying to keep a workforce on the waterfront safe during a pandemic, when folks needed to work in such close quarters while building and repairing ships, promised to be the biggest challenge the yard had ever faced.

After making the first major decision, to implement liberal leave, the yard needed to figure out how to accommodate the changes needed for the remaining workers to complete their jobs, in the yard and out. To meet CDC guidelines, thousands of designers, engineers, and other yard workers who generally worked in offices now worked remotely from home, and they required the IT department to create greater digital accessibility than was ever required previously—access among workers routinely transmitting large digital files. That proved relatively easy compared to the measures the yard took for the steelworkers who absolutely had to enter the yard—after all, you can't Zoom a weld. To get to the waterfront, shipbuilders queued at the gates, and once inside, they needed to access tools, food, and other daily resources, creating other points to gather and wait.

To help develop action plans to address those needs, while keeping the yard operational and Covid-19 compliant, on April 2, yard president Jennifer Boykin established a Covid-19 Crisis Action Group (CAG) of team leaders from all over the yard to work on recommendations on how to safely and efficiently operate during the pandemic. A new vice-presidential steering committee considered recommendations and sent ones it approved to Boykin. This prevented VPs from coming up with individual plans for their own operations and interfering with overall company interests. Four subteams formed under the CAG helped the yard respond to the pandemic quickly and with some agility. The teams met two to three times daily. Each of the teams focused on a different area: mitigating the virus risk to employees; adjusting production to meet customer's needs despite the expected workforce cuts; ensuring yard viability; and protecting job security through and after the pandemic.

Thanks to the CAG process, the yard made quick changes to protect workers, but Covid-19's assault still breached the waterfront gates. What made the virus so insidious was its ability to spread so rapidly once it got into a community—especially into areas packed with people, like a shipyard. The city of Newport News became one of the early spike zones. Everyone talked about creating a bubble, or a series of bubbles, at home, in stores, or at work. But those bubbles popped so easily.

A seasoned shipbuilder, Wilton Ferebee worked as deck electrician specialist and never needed to take a sick day in his life. Never had chickenpox—never even had the flu. Quite an accomplishment for a man of sixty. But when he awoke on Wednesday, April 1, he felt very tired.

"You've been working a lot of overtime," his wife reminded him. "Maybe you need to get some rest."

He said OK and took a vacation day that day. He felt the same on Thursday. In fact, his throat now felt a little scratchy. He didn't want to get anyone sick, so he stayed home again. He stayed home again on Friday. He set up a doctor's appointment for Monday. His temperature, though, started rising on Saturday and his wife gave him ibuprofen to bring it back down. Instead, his fever spiked that day to 103.2°F. His wife called the doctors that weekend, and they told her to keep giving him ibuprofen. If he gets worse, they said, take him to the emergency room.

Ferebee made it through the weekend, and he drove to his doctor's office for his appointment. He pulled up and, as required by doctors' offices throughout the country, called the desk to let the doctor know he awaited instructions in his car in the lot outside. They made sure he had a mask and came out to his car, wearing their own masks, shields, gloves, and jackets. They told him to roll down his window, open his door, and turn in his seat to face them, but remain in the car. One of them pulled out a swab and swabbed his throat and told him to put his mask back on, go home, and stay out of work until he heard from the doctor in three or four days. He drove home, worried.

On Friday morning, the doctor called. "Mr. Ferebee," she asked, "how are you feeling?"

"Not that good," he told her.

She said, "Well, you did test positive. You have Covid-19."

"Wow." Not much else to say at a moment like that.

"Just take it easy," she told him. "Stay in and quarantine for fourteen days, you and your wife."

"OK."

"If there's anything you need, give me a call."

They hung up. Fifteen to thirty minutes later, Ferebee realized he could not smell the bacon his wife was cooking for breakfast. He couldn't taste the cough drop in his mouth, either. A bit later, he got ready to call his boss at the yard, but he could not make the call; he could barely even breathe. He walked to the bedroom to sit in his recliner there and struggled even harder to breathe. His wife came in and asked, "How you feeling?"

He forced out the words: "Can't breathe."

He wound up on the floor, and his wife asked, "Should I call nine-one-one?"

"No, I'm gonna try to get myself together." But he didn't feel any better. She dialed nine-one-one.

As the rescue squad raced to the house, the dispatcher asked, "How's he doing now?"

His wife checked and said, "He's on the floor. He has brown eyes, but I only see the white."

The dispatcher warned, "We're losing him. Tap him on his hand and talk to him." Ferebee's wife put her husband on speaker. "Mr. Ferebee," the dispatcher said. He was unresponsive. "Mr. Ferebee, stay with us."

He heard the siren in the background. They arrived. His wife opened the door and they rushed in. He still lay on the floor, struggling for every breath. The EMTs pushed breathing tubes into his nose, but he started choking and they pulled them out. His heart pounded wildly, and he heard them say, "We got to get him to the truck."

They carried him down the front steps of his home into the back of the ambulance, where they worked on him for thirty minutes. They needed to get blood from him, but the lack of oxygen had caused his veins to start to collapse. They tried both arms. He heard one of them say, "This is no good." He struggled to draw a breath. He figured he was about to die, and he said to himself, *Well, God, I know I'm saved. You did that twenty-nine years ago. But if I've done anybody wrong, please forgive me. I don't want to die like this, but if I got to, I'm ready.*

He heard one of the rescue workers say, "We got one other place we could go but we got to be extremely careful because if we puncture it, he'll bleed out and we won't get him to the hospital in time." It was the main artery in his neck. The EMT felt around to find it and, finally said, "Right there." He pushed it in, and he hit it. He said, "Take him to the hospital now."

At the Sentara CarePlex Hospital in Hampton, the doctors and nurses worked on him in the Intensive Care Unit. They intubated him so he could finally breathe and put him on a ventilator for four days. When they tried to take him off, he still couldn't breathe on his own, and they put him right back on it. The doctors told him, "Mr. Ferebee, we've done all that we can do. The rest is up to God."

He couldn't talk, so he prayed in his inner being: "Lord, help me." He heard a voice say, "Breathe now." He felt his lungs expand. The doctors watched in disbelief.

"Did you see that?" They took him off the ventilator. His oxygen level, which was down to seventy, began to rise, too. He started on the road to recovery—alone. Yes, there doctors and nurses came to see him, but no visitors. His wife became hysterical; she couldn't call him, couldn't talk to him. Ferebee lay there in his room listening to the suddenly ended screams of others. He figured they stopped scream-ing because they died, and the thought got to him.

After being released, he quarantined upstairs in his house for twenty-one days—without seeing his wife. After thirty-five years, they stayed in the same house but without seeing the other in person for three weeks. Being stuck in his bed, staring around the room—the towels stuck under the door to keep his germs from getting out—he felt him-self going nuts. A nurse brought him his breakfast, lunch, and dinner.

Later, he made it a point to tell those in the yard about Covid-19. Devastating. Wash your hands, he told them, and wear your mask, because you can help yourself, as well as another person.

"I want you to picture this," he said. "A little kid, two, three, four, five, six years old. And you come home because you didn't want to wear your mask and they get Covid-19. Can you imagine a child that's two or three years old, can't breathe? And let's say they don't make it. For the rest of your life, you're gonna beat yourself up. Because all I had to do was wear my mask. And I didn't do it."

He wanted to show those who worked with him at Newport News Shipbuilding that this virus was not just a news item happening elsewhere. He wanted them to look at him and say, "I work with that guy." If they did that, he told them, "You will know that it can happen to anyone." They, too, could wind up fighting and praying for their lives.

He wanted them to know that no matter your age—how young, how old—you just gotta take precautionary measures and be safe.

Yard leaders focused on creating precautionary measures for the whole waterfront. The CAG briefed a new shift idea to the VP steering committee on April 9 and then to Boykin the following day. Under the plan, the yard dropped down to two shifts instead of three, to allow for deep cleaning, and then split up the workforce more evenly between those shifts to create better social distancing on the waterfront. The new shifts started near the end of the month.

For those who came to work in the yard, the company installed disinfectant and spray-bottle distribution sites to help people clean their own work areas. The yard also took extra measures to clean bathrooms, turnstiles, and other "common-touch" points. Everyone wore masks. The yard measured and marked six-foot distances wherever workers congregated, such as entrances and tool rooms. Supervisors arranged work to keep trades workers from bunching up. The yard screened all employees at the front gates for high temperatures.

To make the shift structure succeed, the yard had to split not only the production workforce, but also the IT and support staff—the majority of whom traditionally worked the first day shift—as well as the networks, computing systems, and other resources they would need. It would stress people and equipment in a novel way, as the new second shift worked through the night until 2:30 in the morning.

Boykin and Petters offered online messages accessible to all, and the yard president provided answers online to workers' direct questions.

The yard and its HII parent waived the seven-day elimination period for receiving short-term disability (STD) benefits for a bit more than a month, until May 15, due to the pandemic's extended nature. The waiver not only applied to employees who contracted Covid-19, but to all illnesses and injuries normally covered by the STD plan. Workers breathed a sigh of temporary relief.

True to her word, Boykin went online on April 21 to address yard workers directly and honestly. "Dear Shipbuilders," she started. "Although we have not had to close like many other businesses in our community, we are having to redefine how we operate."

Indeed, the working world as waterfront steelworkers knew it began to change dramatically. They had to find a way to work six feet apart all day—getting into the gate, lining up for tools, everything. Everyone looked like a bandit, wearing a mask or bandanna. And the yard had disrupted its entire shift structure, on which waterfront workers had based their family lives. How would that work now, with all the kids being home? Just about everyone had kids at home. Someone needed to stay home with them. At the union hall and throughout the yard, everyone waited to hear what Boykin had to say.

"We recognize the shift change presents childcare concerns for many of our shipbuilders," the yard president told them. *Well, that was an understatement.* "We have been working with the YMCA of the Virginia Peninsulas to find solutions. They offer emergency childcare for essential personnel at several of its centers for children ages five to twelve. In addition, the YMCA will consider accommodating the new schedules of NNS employees if there is sufficient demand for the services. . . . Another option is Child Care Aware, a free state service that helps identify childcare providers still operating in the current climate." *Well, that's something.*

Workers also wondered how they would be able to get to the yard with the new shift schedule. Hampton Roads Transit had already arranged its entire Newport News schedule around feeding the twenty-thousand-worker gorilla's transportation needs. With the shift changes, thousands of workers now commuted during different hours. What did the yard plan to do about that?

"We are actively working with Hampton Roads Transit to provide solutions, and I expect to be able to provide more details very soon about options." *Well, that was of little use.*

"While Liberal Leave (LL) is currently scheduled through May 18, we continue to review and assess the impacts of COVID-19 on the Hampton Roads community and surrounding areas, and make adjustments to LL as needed," Boykin wrote. *Another we-don't-know-yet answer.*

The yard also continued to make, purchase, and distribute masks and other face coverings. "We have fifty thousand on order with a plan to distribute ten thousand per week," Boykin told the workers. "Please remember that these items are in extremely high demand across the world, and while we are doing our best to get them to you, there are many ways to contribute to your own safety and the safety of those around you. I encourage you to make your own face masks from materials you may have at home as recommended by the CDC—not just for work, but for your trips to the grocery store, gas stations, and other places in the community."

She warned the workers: "We will conduct temperature monitoring prior to entry into shipyard facilities as soon as possible. Like face coverings, we placed orders for this high-demand equipment early, but it is taking several weeks to get."

Steelworkers wondered, *If we're such essential workers, why won't Washington get us some thermometers?*

When workers slacked off on the wearing of masks, Boykin wasted no time calling them to account. In her online message of April 23, she addressed them as she always did. Shipbuilders: To be clear, not wearing a face covering when within 6 feet of another person violates NNS Safety Policy.

A fireable offense.

Starting the next day, April 24, Boykin started releasing online the total number of diagnosed individuals who had medically recovered and were cleared to go back to work or to work from home. At the end of April, Mike Butler, who taught an adult Bible study at his church in Suffolk, Virginia, offered the following encouragement to his class:

Historical records indicate that over five hundred million people have died of viruses and diseases across the centuries. This includes the Spanish Flu that killed over fifty million people in 1918. At this point we don't know where Covid will lead us, but we are confident of one thing: God is in total control and in His absolute sovereignty every single molecule is under His authority. We must find comfort in this, so that we, as believers in the gospel of Jesus Christ, can be a calming agent to those around us who are searching for an answer in the confusion and chaos.

Boykin walked the waterfront—channeling, perhaps, the yard-roaming spirit of Homer Ferguson—checking on shipbuilders' progress in making the yard safer and complying with CDC rules. Yard workers filled bottles, delivered supplies, and finished setting up the temperature-taking system the company planned to implement in the coming week. Boykin saw many wearing face masks but stopped and scolded too many without any facial covering, or who broke the six-foot social-distance rule. Some shipbuilders stopped her and complained about coworkers they saw riding elevators maskless. She received e-mails from other shipbuilders expressing concern about the same irresponsibility and selfishness.

The waterfront and the streets surrounding the yard started to resemble scenes from an old Hollywood movie about a mankind-ending plague. With the new shift setup, fewer folks in that first shift meant fewer cars than before; even the parking lots became half empty, as if the vehicles themselves were socially distancing. Fewer people walked the streets—wearing masks—and the streets turned so silent you could hear the pine needles as wind blew them across the sidewalks. The yard also initiated new "no-drive times" when no one was allowed to ride bicycles or operate a motor vehicle inside the gates, to help maintain social distancing.

At sea, meanwhile, all that robbing *Kennedy* to fix *Ford* began to pay off. The navy put its first operational air wing on the ship at the end of May, a major milestone, adding another 1,200 people to the ship's roster. Even better, *Ford* sailors started to use the lower weapons elevators to bring up inert bombs to the flight deck and exercise the system, making more of those problematic lifts operational. The *Ford* racked up more than three thousand takeoffs and recoveries in all, and it had become the navy's carrier flight certification ship for the US East Coast. Maybe now, CVN 79 could get its workers back and, despite the virus, get back on track.

15

A MATTER OF CULTURE

SPRING 2020

Back in the civilian world, the US underwent a world of change thanks to the virus, civil concerns, and politics. Not only did Covid-19 change the way people went about their daily business, but the *politics* of the virus turned folks against one another in a way beyond partisanship—and how anyone could politicize an indiscriminate killer and crippler like this bug was beyond the understanding of most of the shipbuilders. Then, a White police officer in Minneapolis named Derek Chauvin knelt on the neck of a Black man named George Floyd for eight minutes and forty-six seconds on May 25 while Floyd lay handcuffed on the ground in custody, at least two minutes after he stopped breathing—all of this captured on iPhone video, which every media outlet in the country, if not the world, broadcast by that evening. After that, the world outside the shipyard gates seemed to shift into a whole new higher gear of crazy. Around the country—and then the world—city streets erupted in protests led by Black Lives Matter, opposed by counterprotesters of self-appointed "militias."

Addressing the naval community through self-recorded video about the death of George Floyd and the subsequent unrest, CNO

Admiral Michael Gilday on June 3 said, "We need to listen. We have Black Americans in our Navy and in our communities that are in deep pain right now. They are hurting. I've received e-mails, and I know it's not a good situation. I know that for many of them, they may not have somebody to talk to. I ask you to consider reaching out, have a cup of coffee, have lunch, and just listen."

Looters, rioters, and those of a similar ilk took advantage of the cultural chaos to pillage, destroy, and disrupt. Hampton Roads did not escape. Newport News police chief Steve Drew even marched with protesters in Newport News while another, rowdier group closed Interstate 64 near the tunnels. The nadir came about two hundred miles to the north on June 1 in a moment captured again on film by every major global news outlet when federal forces cleared a path from the park to St. John's Church—known as the Presidents' Church because previous presidents went there to pray—and the adjacent streets, teargassing and cudgeling peaceful protesters out of the way like a Greek phalanx for Defense Secretary Mark Esper, Joint Chiefs of Staff Chairman General Mark Milley, and other federal henchmen to accompany Trump to the iconic house of worship.

The steelworkers strove to keep the chaos outside the gates. Local 8888 president Spivey addressed the waterfront in the June edition of the *Voyager*:

There is an unrest in our communities and nation that is visible to us all. I trust that we make positive decisions when addressing your opinions. Think, does it show respect and concern towards others who may be affected? This country is a great nation and all are great people who have great love for one another in some regard. Please express yourselves with respect. Don't let your expression divide us as people any further than we are . . . brothers and sisters, pray for one another and show some love. We're all in this together.

The same day Trump's military muscle marched on St. John's Church, in Newport News a small group of sailors and officers assigned to the carrier *John F. Kennedy* nuclear spaces made their way along Pier 3 on the shipyard waterfront, enjoying the nice spring weather as much as possible, given the circumstances. They wore their tan plank-owner caps, cammies, and of course, masks. They became the first sailors to retake CVN 79 following the Covid-19 retreat. They returned to work and joined the steelworkers on the ship finishing the *Kennedy* spaces. Spivey left the union hall and made his way to the front gates to show his support for the steelworkers and answer any questions—wearing a mask, of course.

Local 8888 passed out union face masks at the end of June between 8:00 a.m. and 6:00 p.m. every Wednesday and Thursday—two to each steelworker member, one reusable cloth mask and one disposable paper mask. Virginia started to reopen, relaxing some shutdown measures, but Spivey and other local leaders asked members to continue to take precautions. He appeared at the gates before shift starts, wearing a mask, talking to the members, and answering any questions. The union hall also resumed full operations again, requiring all members to wear masks in the building.

Back inside the yard, Boykin kept up her patrols, but the virus remained relentless. On June 15 alone, the yard reported five new cases—including four workers aboard the carrier *Washington* at Outfitting Berth 1, bringing the total number of cases among those working on that ship to thirty-one since April 1, including twenty-three confirmed in June.

The yard's cleaning crews stepped up their efforts, and the CAG considered hiring more outside professional disinfecting services for cleaning trailers, trades buildings, shipping containers, restrooms, and other common-use areas, as well as increasing air flow and ventilation in targeted work areas, closing communal kitchen areas, providing more hand sanitizing stations and products, and broadening contact tracing. The yard collaborated more and more with the navy and its medical teams. Covid-19 is not gone and we must remain vigilant here at work and through our personal decisions outside of work, Boykin warned online.

Despite Covid-19's impact, Marzano continued to build up his crew on the *Kennedy*. Indeed, when former *Ford* CO Rear Admiral John Meier, who became commander of the Naval Air Force, Atlantic, visited the ship and its commanding officer in mid-June, he was surprised that the *Kennedy* was outpacing the *Ford* for number of boarded sailors at the same stage of construction. He retained a particular interest in keeping *Kennedy* work on course. His mother had gone to Kennedy's presidential inauguration, apparently pregnant with the future Rear Admiral Meier. As a kid he cherished a large PT-109 model, complete with battery-powered propellers. The Kennedy legacy steered him into the navy—and guided his life. Kennedy's civil rights stance weighed on his mind during the recent unrest in America. At the same time, he studied the new carrier *Kennedy* and thought about how this ship underscored another of JFK's basic principles: pray for peace but prepare for war.

Meier noted that the *Kennedy* was no longer a construction project, as the yard turned it over, system by system, to the crew, as it became a true navy ship. How it differed from buying a car—taking a complete car off the lot for a test drive and buying the whole thing at once. Instead, it would be like buying a car's engine first, then its radio, then its steering wheel. . . . The sooner more sailors climbed aboard, the sooner those sailors would imbue the ship with their character and culture—as molded by Cherry Marzano. Meier conveyed how impressed he was with the command culture the *Kennedy* CO had established.

"You've got this heritage and culture, and the namesake of the ship, and the unique opportunity to build a culture, and a command climate," he told Marzano. "That is the art of command."

While *Kennedy* sailors had already shortened training time by getting on the *Ford* for firsthand operational experience, Covid-related impacts still threatened to crimp the overall CVN 79 schedule. Like most, Meier found it impossible to accurately gauge the long-term effect of Covid-19. It depended on when all the steelworkers could get back to the waterfront.

Despite what Rear Admiral Meier thought about the construction status of the carrier *Kennedy*, the yard understood there remained a significant amount of work to be done, particularly due to Covid-19 delays. True, the waterfront had reduced virus-related hurdles, thanks to the leadership of Jennifer Boykin—Mike Butler was impressed and inspired by the way she personally embraced the challenge, the transparency of her status-update e-mails, and her generally upbeat nature during all if it. "She made it a personal challenge," he recalled. That made a difference on the waterfront, he noted. "We came through this with an attitude of, 'OK, what's next?' If something else unexpected falls on our lap tomorrow, it's OK. We know how to deal with it."

Boykin got down on the waterfront with the rest of them, setting the example, wearing a mask, maintaining distance, and getting her temperature checked. She made it clear from the very beginning: "We have to look at the personal aspects of this, of every person's life, to understand what Covid is doing to their lives and figuring out how to help them out."

Butler certainly got that. The question remained: How do you manage virus protection in a city 1,200 feet long and 275 feet wide filled with 2,500 people? How do you keep Covid from spreading throughout that small city? How do you do that *and* maintain a schedule?

Initially, Butler planned to keep his teams out of ship spaces near where the nukes expected to train, but the navy curtailed that training to protect the force against Covid. Where would the navy need spaces next? Where should the CVN 79 team put its workers? Whatever the answers, shipbuilders needed to make the *Kennedy* a safer place to work and reduce the threat of infection. To do that, the yard installed additional, temporary sinks and water fountains for workers. As the warmer weather hit, the yard also quickened the startup of air-conditioning units—to keep masked workers cooler. In previous summers, the yard had made sharable and refillable large water coolers—as well as the bags of ice—available to help combat the heat and keep workers hydrated. Those proved unsafe in the current Covid climate, so instead the company provided personal one-gallon water jugs, making workers responsible for cleaning their own jugs at home. The company also provided seven-pound ice bags to workers, but not for sharing. Even though liberal leave ended in mid-May, thousands of people still worked from home, and Butler needed more workers. "We're still at the height of it," he and other yard supervisors reminded those on the waterfront.

Nuclear-division workers Warren and Sheila Outlaw found out just how right Butler was.

Like so many others in the yard, they followed the rules to stay safe and healthy. Warren, as a supervisor, not only did everything as he was supposed to—wearing a mask, washing his hands, and keeping at least six feet apart from others—but he constantly reminded his workers to do the same and made sure that, under his watch, they followed all the guidelines, not only in the yard, but also when they went home to their families.

The Outlaws, though, discovered that a huge difference existed between your *real* family and the bubble around it and your *extended* family, lifelong friends who've become like family. You never knew what kind of safe family bubble they established, and that fact was made clear to the Outlaws that spring. Sheila's childhood friend from Newport News, whose granddaughter had just graduated high school, decided to hold a party, just like thousands of other questionable graduation parties held that June in 2020. The Outlaws attended the celebration with their second family.

A couple of days after the party, Warren did not feel right.

He thought maybe his diabetes was throwing his body out of whack, so he decided to take it easy. He never liked to make a fuss when he got sick, but when the following day brought no changes, Sheila persuaded him to take a Covid-19 test. Awaiting the results—in most parts of Virginia then it took more than a week—he moved about the house like an old man and struggled to catch his breath. Sheila called his brother-in-law, who lived in Kentucky, and he persuaded Warren to go to the ER. Sheila called for an ambulance.

The EMTs arrived and began to question him. "What day is it?" He looked at them and shrugged his shoulders. He understood exactly what they asked. *He asked the day—so just tell him what day it is.* He knew—but he couldn't say. Something wasn't right, and it showed. The one EMT glanced at the other one and said, "We got to go—now."

Warren started to walk with the EMTs out of his house, heading for the front door as he had done thousands of times before. For some reason, he simply failed to navigate his way to the door. He walked into the wall instead. *What is going on?* He never did that before. The EMT behind Warren grabbed the shipbuilder by the shoulder to steady him and guide him through the door, into the ambulance, and to the hospital ER, where they told him he had pneumonia. After that, most of the communication between husband and wife went through hospital staff, who told her that, judging from the X-rays, he likely had Covid-19.

At first, Sheila and Warren talked to each other on his cell phone, but every time she called, he struggled harder to talk. Warren texted her they were about to put him on a ventilator. Right after, the doctors called to say they had put him on ventilator and placed him on his stomach to help him breathe. He had been in the hospital two days. Warren became delirious and he hallucinated. Sheila filled him in when he came out of it days later.

They tried to take him off the ventilator, but he fought to breathe. Sheila prepared for the worst. The family prayed hard, as did those at

Macedonia Baptist Church. It surprised Sheila that so many people cared about them in the community and in the yard. Her supervisors, even the director of her department, all reached out to her. They called every day. The vice president checked in on her. The phone rang and dinged with texts around the clock. It created even more stress; she stopped answering the phone and the messages and sought out a yard-provided therapist, who helped, a little. What she needed was to talk to Warren, to hold his hand. She kept herself together, for her family's sake, she thought, as she sat alone in the home she and Warren had made together.

Finally, after another couple of days, Warren breathed without the ventilator. He tried to talk to the nurses and doctors, but no sound came out, at least nothing like the true sound of his own voice—one of the side effects from the ventilator. Every time he tried to speak, they gave him a look that asked, "What are you trying to say?" That frustrated him. It took him a week to speak at an audible level. Not that he had enough energy to talk, or that he had energy to do anything. He tried lifting his cell phone, but it seemed so heavy. It took too much out of him to even pick it up. He willed himself to lift the phone and hold it in front of him to call Sheila. His password eluded his memory. *What is going on? Why can't I remember my password?* The phone must be damaged, he thought, not his mind. He told his nurse, "My phone is not working, what happened? Did I drop it into water?"

He finally got the password but needed the nurse to dial the phone for him. His hands, his fingers, no longer obeyed him. He finally called Sheila, who told him she had prepared for the worst. He faced a long journey to recovery, though. The nurses strapped him to his bed to keep him from falling and hurting himself. His sense of balance remained wobbly for a long time.

Seeing Warren breathe on his own lifted a great weight off Sheila. He wanted to go home now that he could breathe, and she now worried about her husband's desire to leave the hospital just because they took him off the ventilator, even though he needed help just to stand up and struggle to the bathroom. From being on the ventilator and lying in bed for so long, his legs hung below his waist, useless. When he tried to stand, they revolted with a clear message: *I don't know where you're going, but* we *ain't going nowhere.* His nurses told him, "Just take your time," but he watched the news shows and saw people being stuck in the hospital for three months, and he had no intention of lying in bed one second longer than necessary. "You're trying to come home too early," Sheila told him. The doctors took blood all the time—he dreaded the sight of needles for the first time in his life—and checked his breathing. They planned to keep him until they found him ready to go.

After about a week, he finally went home—to a nightmare. He needed two canes to stand. Sheila wrestled his bearlike frame around the house, a bit easier to do now that Covid had robbed him of twenty pounds. But they quarantined from each other for fourteen days. She ran in, helped, and ran out. The quarantine ended, and he grew stronger, gradually. But he still often lost his breath and his balance. He never regained his old strength. He learned to take his time with everything he did.

Every time he saw folks maskless out on the street or on the TV he shook his head. The Outlaws made their story known around the shipyard, even releasing an online video describing their ordeal. They wanted people to protect themselves without learning the hard way just how vulnerable anyone could be to Covid-19. They hoped everyone took the virus seriously.

Some news items managed to distract public attention from Covid-19 for a short while. Global media reported that Russia offered bounty payments for killing American soldiers in Afghanistan. Military officials worked to clear public spaces and work areas of Confederate flags to match the national mood on the subject. Closer to home, for

Newport News, anyway, the navy fired the head of the carrier *Ford* program. Folks ran out of patience awaiting fixes for the elevators and other ship systems.

Technical issues regarding the elevators and other systems now delayed the first deployment to 2022 at the earliest, or about five years after the ship was commissioned into service. Worse, for those on CVN 79, Covid-19 and the *Ford* fixes continued to delay *Kennedy* work. During normal times, the yard ramped up the workforce to catch up. The virus, though, prevented anything like that.

Like the Outlaws, Jennifer Boykin hoped everyone in the region would take Covid-19 seriously as the virus refused to wane in the summer months. In a July 14 message to yard workers, she noted not only a rise in community cases, but also some "social-distancing lapses during eating breaks inside the yard." As a result, she lamented, the yard saw the highest spikes in the region and the facility thus far in July.

Boykin also noted that Virginia governor Ralph Northam reported a troubling increase of Covid-19 cases in Hampton Roads. As of mid-July, the seven-day average of new cases in the region was 356, compared to about 60 in early June. The incidence of positive cases in twenty- to twenty-nine-year-olds rose 250 percent in Hampton Roads since early June.

The governor attributed the new-case-average increase to substantial community spread from people socializing without wearing masks, especially young people.

"What happens outside our gates impacts what happens inside them," Boykin warned. "This is why we urge you to practice social distancing and wear a mask whenever you are in public."

She added, "Our cases will very likely continue to grow. We are in the process now of refining our long-term operations plan to address how we safely operate with no foreseeable end to Covid-19." In the midst of this, Virginia became the first state to issue work safety standards for coronavirus—not the best environment for building a new aircraft carrier. For some, like supervisor Don Doverspike, tougher work rules created greater headaches.

After the ship launch, Doverspike began to coordinate the building of the ventilation system—no small task for a small mobile city. Covid-19, though, forced him to work from his home in nearby Williamsburg. He had an immune deficiency, and the recent spikes made it impossible for him to go to work when any workers were there. Sometimes, he headed in between shifts, with no one else about, to inspect the work and leave before the next shift arrived. He grew to like his empty-ship patrols. He made himself a plan, a "honey-do" checklist, of what he needed to look at—which compartments, the status on each, and so on. He missed the interaction with people—he fancied himself a face-to-face guy, but such interactions had to wait, probably for a while. He missed those moments of "Hey, have you tried it this way? Or "That's a different way to look at it." So now he relied on e-mails, texts, and phone calls. He surprised himself how much work he had been able to complete.

However, completing work on the *Kennedy* became more difficult for Doverspike, Butler, and everybody else on July 20 when a fire broke out on the ship. Some hotwork sparks caught flame belowdecks, and the fire watch doused it. No one dismissed a fire, no matter how small, on a ship. Every errant spark meant potential disaster. Steelworkers stopped all hotwork on the carrier and the yard evacuated all the yard workers, sailors, and officers. The fire aboard CVN 79 never came close to the smoky blaze at the San Diego navy base that totaled the amphibious assault ship USS *Bonhomme Richard* earlier that same month, but it was enough to halt hotwork on the *Kennedy* for two days. After a hurricane, the resource stealing for *Ford*, and the pandemic-related workforce shortages, it now seemed painfully obvious that *Kennedy* work would never meet schedule—even if the yard and navy would not say so—and who knew what would happen with those man-hour reduction tallies, with the ship about three-quarters complete.

The yard and navy, though, grabbed for a lifeline. Recall that deal to deliver the ship in two phases. Assistant Secretary Geurts always hated that plan; he thought, when you got a ship, you got the whole ship. To meet that first-phase delivery date, shipbuilders struggled to make up lost time, racking up the man-hour totals. But if they eliminated that first-phase due date of 2022 and instead agreed to deliver the whole ship somewhere near September 2024, they could stretch out the work and take some of the pressure off. Also, truth be told, such a change made it easier for *Kennedy* Reactor Officer Captain Klopfer to complete his training schedule. The yard and navy started to negotiate a single-phase delivery for the entire carrier just as Newport News began moving metal around for three carriers in its yard—the *Kennedy*, the *Enterprise*, and the *Doris Miller*. The last time the yard built three carriers at one time, the first President Bush sat in office.

The question continued to be how much work steelworkers could complete on any carrier as Covid-19 ravaged the workforce. On July 21, the yard reported thirty new cases inside the gates, with more being exposed to one another during lunchtime. Boykin told employees, "I implore everyone to be conscious about how closely you sit to others during lunch." Issuing his own warning to steelworkers that the virus hadn't gone anywhere yet, Local 8888 president Spivey spoke with yard president Boykin about more ways to protect the steelworkers.

It became equally difficult to keep some of the cultural and political chaos out of the yard. On the cover of its July edition of *Voyager*, Local 8888 splashed pictures of Biden and Spivey together, taken when Biden had visited the yard in the fall of 2018—promoting a virtual round table taking place that month in 2020 on Biden's "Build Back Better" campaign to revitalize US manufacturing, somewhat reminiscent of the 1920 Warren G. Harding campaign, promising a "return to normalcy."

While the rest of the nation was obsessed with the Black-versus-White conflict, steelworkers focused on their everlasting battle of labor versus big business, and they presented Biden to the membership through that lens. In the August issue, Local 8888 printed this message from Gene E. Magruder of X33, chairman of the political action committee:

> Joe Biden grew up in an industrial town. He grew up knowing workers and their issues, and thus over his thirty-six-year career in the Congress voted in workers' favor eighty-six percent of the time. While Trump's career has been consistently anti-labor, from his days of ripping off contractors, daring them to take him to court and forcing them to accept way less than the signed contract. He has attacked workers' rights since becoming president and has unequivocally said that American workers need to take a pay cut to compete.

Tidewater provided steelworkers with another reminder of its hellish 2020 summer, as temperatures reached over 100 for several days in August. The political climate heated up inside the gates as well. In late August, Dan Sunderland wore his TRUMP 2020 baseball cap at work—just one of the several pro-Trump caps he wore every day after Trump started running the first time. This time, one of the foremen decided to make an issue of it. Not his foreman; in fact, his direct foreman complained only once during the past four years about a cap he wore—a blue one that said, MAKE AMERICA A SHITHOLE. VOTE DEMOCRAT. The supervisor found that language offensive, so Sunderland stopped wearing that one to work. He wore the other caps, and some complimented him on them. The caps sparked conversations. That's what people did at work, they talked about things, even their differences.

Therefore he thought nothing about wearing his black cap with red-white-and-blue trim and white TRUMP 2020 lettering from the parking lot to the little two-story shack his team had their safety meetings in, just as he had over the past four years. After leaving the shack, and before entering the yard, he traded the cap for his company hard hat.

He walked into the shack, and a Black foreman, someone he had never seen before, told him to remove the cap because of shipyard rules against wearing campaign items. Sunderland didn't know about that, but he did know of one shipyard policy—you had only one boss in the yard, and this guy wasn't his. He ignored the excited man in front of him. If the guy had a problem, he could take it up with Sunderland's direct foreman. That's exactly what happened.

Sunderland got into the notoriously slow elevator. The enraged foreman bolted up the stairs and was talking to the Trump-supporting pipefitter's foreman by the time Sunderland got to the second floor and sat down at a table to await whatever was about to go down.

His foreman told him to remove that hat.

As far as he was concerned, the First Amendment guaranteed his right to wear his cap. This might be a private company, but they built carriers and subs for the US Navy. He said no.

She explained that company policy forbade anyone from wearing political clothing like that. He remained adamant. "You can and will be fired," she warned him.

The answer remained no.

He would be fired for refusing her direct order to remove the hat.

That's silly, he thought. That rule was really a safety thing. If a supervisor warned you not to do something or you could die or some such thing and you refused. This was different. She warned him she'd call security to remove him. He kept the cap on. She called security, and the guard escorted Sunderland out of the yard.

After being suspended for three days, Sunderland was told to report to the yard's human resources office. A yard official there again reminded him of the policy against campaigning in the yard, then fired him for "refusing to follow a supervisor's instructions."

What campaigning? He passed out no bumper stickers. He asked no one to vote a certain way. He did nothing but wear a ball cap going to work.

Even though Sunderland had paid no dues to Local 8888 for about a year, the union still sent a rep to quietly sit in on the meeting between the worker and the HR official. The local rep wanted to ask the yard to give Sunderland another chance but was told the Trump supporter didn't want that. Spivey never wanted to see anybody fired over these kinds of things, and the yard had been cracking down—supervisors told another steelworker to turn a Black Lives Matter shirt inside-out.

As for Sunderland, the fifty-five-year-old had no regrets about sticking to his guns. His story made the national news. He was featured on Fox and even wound up talking to the White House. Spivey wanted to file a grievance and possibly get Sunderland's job back, but the union didn't hear back from him on those offers either.

Despite the experience, Sunderland didn't consider the yard to be racist. Everybody generally got along. His crew was half White, half Black, and even after he got fired, he helped a former Black workmate, a friend, carry some things to the dump in the back of his truck. But to Sunderland, it seemed all that inclusion and diversity "stuff" only divided folks. "Why have a Black History Month?" Sunderland wondered aloud when asked about the incident later. "Why not just have history all year long—American history?" At the yard, they started an Employee Resource Group, for minorities. "Why have something like that if we're all the same? Why are there extra points if you're a minority or a homosexual to get a job? The only thing that should matter should be your qualifications, not your race, not your background."

He failed to grasp one of the blind spots people had regarding Trump, especially in Newport News. Trump increased defense spending. Obama reduced it. He was unmoved by the fact that, although spending *had* been frozen during most of Obama's time in office, that had been largely due to a Congressional law, not his administration's desires. As far as Sunderland was concerned, Biden promised to be another Obama, even though the Democratic presidential candidate had publicly said in September that year his plans included no defense cuts.

Nationally and internationally the focus sharpened around defense spending and the need, despite continued dire economic straits due to Covid-19, for more military money. Such a mind-set usually bodes well for Newport News Shipbuilding, particularly when the impetus for the concern during that early fall of 2020 became China and its shipbuilding programs, as well as Chinese missiles developed to take out US carriers. China test-fired two missiles in September into the South China Sea, including a "carrier killer" DF-21D rocket into an area between the southern island province of Hainan and the Paracel Islands—a notable feat, although much easier than actually hitting a moving, protected carrier, an ability China has yet to prove. Still, China apparently recently built a stockpile of two hundred DF-26 missiles, compared to about sixteen as of three years earlier.

Later that same month, the Pentagon released its annual China report, noting the leaps the Chinese military and associated industries made. For example, China now ranked as the top ship-producing nation in the world by tonnage, having increased its shipbuilding capacity and capability for all naval classes.

As previously noted, China had commissioned its first domestically built aircraft carrier in late 2019, and the Pentagon reported that China expected its second domestically built aircraft carrier to enter service by 2023, a year after Newport News planned to deliver the *Kennedy*. The Pentagon report stated: "China continued work on its second domestically built aircraft carrier in 2019, which will be larger and fitted with a catapult launch system. This design will enable it to support additional fighter aircraft, fixed-wing early-warning aircraft, and more rapid flight operations and thus extend the reach and effectiveness of its carrier-based strike aircraft. The PRC's second domestically built carrier is projected to be operational by 2024, with additional carriers to follow."

According to the Pentagon, China focused on protecting its maritime approaches and threatening regional opponents, using its slowly growing blue-water fleet to protect overseas holdings and sea lanes. In addition, however, it continued to prepare for a much bigger military challenge against one particular foe: the US. As one analyst put it, "It's like a football team who's certain they're going to reach the Super Bowl against a certain team and is building a team just to beat that team, but they're building a team that's not planning to reach the Super Bowl for a few more seasons."

It had been the Trump administration that had first truly identified China as a major military foe and tailored the Pentagon resources, particularly those in the navy, to battle Chinese forces. Trump's 2016 campaign platform put a bigger navy front and center. So, Trump's Pentagon China report, released in the middle of the 2020 campaign season, should have been dry kindling for Trump as he made his stops, including one at the Newport News/Williamsburg International Airport on a windy, rain-soaked September 25 evening, at the former site of a hospital for the chronically ill. A throng of thousands—some who'd waited through the wet weather since dawn for Air Force One to pull up to the tarmac—didn't hear about any of that or know that the nation's largest naval shipbuilder existed just down the road. Nor did they hear any concerns about the Covid-19 pandemic sufferers. They, instead, cheered the president's fearful warnings about the fate of the nation under Democrats and the progressive left.

Avoiding such cheering throngs, Covid-19 survivors Sheila and Warren Outlaw remained indoors just a few miles away from the airport. He had been recuperating through the summer, but he still struggled to get around the house, let alone get outside. He took it one day at a time. He missed working and being around people on the waterfront. The concern that everyone at the yard still showed for him and his wife truly humbled him, just as he and Sheila remained dumbfounded by those who continued to deny the existence of the virus or its impacts. They shook their heads when they saw people without masks in public. People seemed to think of Covid-19 as just another type of flu—catch it, quarantine, and get on with your life. They experienced firsthand how such an untrue mind-set threatened everyone. They lost a couple of friends and a family member to Covid-19 over the summer. Warren still got winded and endured dizzy spells, but he survived.

Like Warren Outlaw, other steelworkers kept away from the yard as Covid-19 continued to ravage the region, the state, and the country, and not only because they worried about the virus. The waterfront was a dangerous, tough place to work. Workers still got hurt—workers like Jordan Patterson, who had bent his body one too many times the wrong way to finish a paint job. He wrenched his back and the yard restricted him to sedentary duties. No such work existed in his department, and he might no longer have a job.

Elsewhere, Butler and his teams worked full throttle on the *Kennedy* with the workers they had while the navy and the yard negotiators hammered away at a single-phase delivery plan meant to save the schedule, as the US election wound down to a finish. Then came another shot across the navy carrier plan's bow—a new Pentagon plan, called Battle Force 2045, which promised, yet again, to cut the future carrier force.

Just before the election, Defense Secretary Esper released his plan to slice into the future carrier fleet size, equipping the fleet instead with smaller light carriers. Again, department carrier killers dismissed the big ships as dinosaurs of the future, calling for all that money to be shifted to a bunch of robot ships on and below the sea. The new Pentagon placed all that work to develop a more affordable, serial-production Ford-class ship—starting with the *Kennedy*—into jeopardy. The plan threatened to slow carrier production, which also jeopardized all the carrier-construction savings negotiated by Hondo Geurts's and Mike Petters's teams, while making it impossible for the nation to ever achieve a force of twelve supercarriers.

Geurts remained optimistic about the two-carrier deal, though. "The 2045 plan is no attempt to undo the two-carrier buy," he told reporters at the time. "Nothing I am aware of would impact contract and delivery of these two carriers."

Still, it particularly galled carrier proponents that every wargame being used to kill the carriers in DC kicked off at what they call Phase Three—with bullets and missiles already flying. The games featured nothing to highlight the carriers' importance in Phases Zero to Two, when a carrier's very presence could prevent the start of hostilities; no thought was given to the carriers' message-warfare value. People needed to think about carrier presence the same way they perceived nuclear deterrence. Jobs in Newport News and Tidewater often depended, not on the ability of steelworkers to perform their craft, but on the theoretical whims of analysts manipulating spreadsheets instead of sheet metal.

The election of Joseph Biden and the subsequent resignation of Secretary Esper put the Battle Force 2045 plans on hold. Shortly after the election, the navy released its thirty-year shipbuilding plan, which retained the same number of carriers over the coming decades, but also included the new light carrier development, without detailing how the service expected to pay for both—something that no analyst thought feasible. Biden's win, by about 81 million to 74 million in the popular vote, made him the second Roman Catholic president, after John F. Kennedy, and it put the country on a different political course, recalling JFK's words in *Why England Slept*: "For an election is, after all, the best barometer of popular will in democracy."

In the Newport News Shipbuilding yard that election day, on the *Kennedy*, Captain Klopfer's nukes received an education in nuclear physics in a hybrid virtual/live lesson, meant to maintain social distancing. It was, of course, classified, and few outside that group would have understood it anyway. A sailor at the entrance to the space checked everyone's temperature with a little "gun" tap to the forehead. This area belonged to the navy and the nukes, and they protected their turf. Elsewhere on the ship, the walkways and spaces looked like chaotic construction zones, with wires, piping, and venting systems strewn about and leading this way and that. Steelworkers scurried about, many with tablets. The air tasted of metal, dust, and stress. The din, as always, deafened—the sound of a carrier coming to life. Butler and Cherry Marzano made their rounds together on the ship, another sign of yard and navy folks working joined at the hip to get it done.

A palpable sense of relief hung about the ship. The yard and navy had agreed on a single-phase delivery, and the shipbuilders and their navy partners breathed easier. The yard was to deliver *Kennedy* as a finished ship about three months earlier under the new plan, in the third quarter of 2024. The new delivery plan included a new jet-blast deflector, up-dated communications equipment, and other modifications needed to handle the new stealthy F-35 Joint Strike Fighters. Again, the navy relied on shipbuilders to redesign and integrate all that. The changes erased any doubt about a higher construction price tag for *Kennedy*, but as Con-gress had required the changes, lawmakers had little choice but to accept the additional cost. Exactly how much more remained unknown, as the yard had yet to reckon the impacts of the raging pandemic. Mike Petters told Wall Street analysts the new delivery date reduced risk on the *Ken-nedy*, especially on testing, the phase in which they wound up having their major problems on *Ford*. "We're resetting the test programme on that ship over the next couple of years," Petters told investment analysts. "It's a chance to reset the sequencing," he explained. The total price now, including navy systems, had reached about $13.4 billion—again, though, without accounting for Covid-related costs.

Kennedy steelworkers now focused on large shipwide systems— air-conditioning, water, ventilation, electrical, and sewage. They needed to complete six hundred ventilation systems alone. Just be-low the flight deck, steelworkers hooked up some of the fixtures to the landing-gear turbine, which resembled a giant water heater. They muscled and coerced the giant twister machinery through a hole in the decking. At the same time, they installed EMALS on the deck, under a long, metal sloped covering, as done earlier on the *Ford*. Also, on deck, steelworkers finished the fittings and openings at the top of the island—like a huge inverted French window—for the new radar.

They restarted work on the elevators using talent from the *Ford* team. About two hundred steelworkers who accompanied CVN 78 out at sea, building and testing the lifts, boarded *Kennedy* when they returned, to impart their knowledge. No one talked man-hours anymore—a welcome change after more than half a decade focusing on those numbers.

On a day like this, looking out over the *Kennedy*, apprentice electrician Matt Phoebus tried to wrap his head around the sheer feat of building a ship like this. *We built this. We put this entire thing together. People like me built this.* Not even a pandemic could stop them. The navy needed them. The yard took care of them, and they took care of everything else on the carrier.

He watched other shipbuilders scrambling around him, some with stickers on their hardhats that proclaimed *WE BUILD FREEDOM.* It was something to take pride in. And they did. They got to show up for work every day and make a difference.

That sense of pride swelled union membership. Spivey surpassed his ten-thousand goal and even registered more than eleven thousand. Bolstered by its expanding force, Local 8888 rejected the yard's proposed new union contract, demanding more across the board. Back in the yard shops, robots and steelworkers were already cutting and welding steel for the *Doris Miller*, while out near the dry dock, Charlie Holloway guided Big Blue to lift and flip a two-hundred-ton pancake of steel plating for an assembly that would become part of the *Enterprise*, the next carrier. Out in the distance, the *James* shimmered in the sunny, seasonably chilled November air. Swiveling around, the crane operator surveyed the yard—the new shops, the old shops, *Kennedy* rising next to the distant pier, and the streets filling with trucks coming down from Pennsylvania off Route 664, full of steel for future carriers, headed for a place that was once known as Point Hope.

16

SHIP SHAPE

SPRING 2021

Over the previous decade, shipbuilders and sailors had stiffened the *Kennedy*'s spine with steel and infused its spirit with the navy culture and character. They now needed to get that heart beating and its systems circulating all the electricity, water, and air to make CVN 79 a living, breathing, fighting warship, an effort that would take another couple of years as the waterfront workers started up, operated, and tested the millions of miles of cabling, piping, and other arteries that made it possible for the ship and its thousands of human partners to coexist in the most challenging environments on the planet. While the rest of Tidewater escaped a brutal heat wave by watching Olympic Games coverage out of Japan, steelworkers and navy engineers worked literally side by side now, turning on major electrical, ventilation, and cooling systems, running them and inspecting them, trying to find any problems now rather than years later in combat at sea.

Commander Angela Owens, the *Kennedy* chief engineer—CHENG, in navy parlance—knew there would never be a better time to do such work than at this moment.

True, with a team of sixty-five at the moment, she had only a bit more than a quarter of her total engineering force, which itself, like other Ford-class ships, operated with a third of the engineering team compared to the force on Nimitz carriers. Design dictated the reduction—more sensors and automated systems meant fewer folks in engineering. Being a no-nonsense product of Texas, she easily managed to tackle major projects with little more than a shrug.

With the adoption of the single-phase delivery, the *Kennedy* CHENG expected to take over the systems in early 2022, but to accomplish that turnover, the shipyard's testing teams and the carrier's engineering team needed to start making those transitions now as the yard continued to turn over some of the spaces to the navy.

By the summer of 2021, CHENG Owens's engineers claimed ownership of about six hundred spaces, about a fifth of the total. After years of doing their own thing, separately, shipbuilders and sailors started to form a joint force of sorts. The *Kennedy* engineers walked the systems every week, while the shipbuilders showed the sailors the ropes—well, really the cables, wires, pipes, and so on. With the Delta variant causing Covid spikes throughout the state, everyone wore masks again. Vaccine shots, though, remained more elusive among the steelworker tribes.

About half the waterfront saw it as a matter of basic freedom; they simply didn't want the government, the company, or anyone else in authority to rule over the care of their own bodies in that way. Those feelings ran across generations, from master shipbuilders down to apprentices—even the apprentice baseball team put its season in jeopardy because so many players refused to get vaccinated.

Boykin and her team faced questions again about the workforce. What if this new Delta variant breached the gates? What if it decimated the workforce in a way the first waves did not? They imagined the impact on the carrier-building schedule, not only for the *Kennedy*, *Enterprise*, or *Doris Miller*, but also for the carriers beyond. Valve, pipe, and steel manufacturers around the country already dreamed of— and pushed for—another CVN block buy, and any kind of schedule disruption threatened to abort such plans before they had a chance to grow in the budgetary womb of the Pentagon.

During those first few months of the Biden administration, that womb proved to be somewhat infertile. With the trillions being spent, promised, and proposed for Covid relief and infrastructure, precious little remained for defense. The navy's requested increase failed to even cover inflation, though it covered all immediate carrier needs.

Petters, Boykin, and others in Newport News appreciated the support but also recognized the uncertainty of the future when the service released its long-term shipbuilding plan. Unlike the Battle Force 2045 plan under Trump and Esper, with its very definitely defined large naval buildup, the Biden offering included an incredibly broad range of potential ship counts for the future fleet. Although the plan dropped the idea of a new smaller carrier ship, the plan's "range" for a carrier fleet ran from a low of nine to a high of eleven—never reaching the Congressionally desired dozen-ship fleet and possibly dropping below the eleven mandated by law.

Usually, the White House releases its annual budget request in the first couple of months of the year, but everyone expects a new administration to delay a bit. Still, Biden's holdout until May 28 irked Congress. The details of the budget, particularly the proposed funding for the navy, irked lawmakers even more. The fiscal year 2022 Pentagon budget proposal requested nearly $212 billion for the Navy Department, which includes the Marine Corps, an increase of almost $4 billion, or 1.8 percent compared to the navy's enacted budget in fiscal 2021, less than enough to cover inflation. The proposed fiscal 2022 shipbuilding account of almost $23 billion represented a 3 percent drop from the enacted 2021 amount.

The real purchasing power of navy funding had remained essentially flat since fiscal 2010, and House Armed Service Committee (HASC) members railed against the cheap budget proposal, the navy's wishy-washy shipbuilding plan, and the navy's inability to keep pace with China, the leading US threat.

Virginia Congresswoman Elaine Luria, HASC vice chair, noted that during the current century, the US Navy's fleet advantage over China of 75 ships shifted to a disadvantage of 37. "That's a swing of 112 ships," she said during the June 17 HASC subcommittee hearing, asking, "Are we keeping up with them today?" In response, Vice Admiral James Kilby, deputy chief of naval operations for warfighting requirements and capabilities, replied, "They are our pacing threat; we're not keeping up with them." Luria asked what the navy is offering in the proposed budget to counter the current threat. "I don't see what the navy is doing today to accomplish that."

The budget, she said, appeared to be based on hope for technology the navy might get in the future. That sense of vague hope, HASC members complained, was further reflected in the navy's thirty-year shipbuilding plan, published on June 17. The plan offered such a wide range of possible fleet outcomes that lawmakers could identify no examples of the service overstating or understating its future force. While the plan echoed some of the fleet recommendations reached in Pentagon studies and released in 2020, both US lawmakers and navy leadership acknowledged that the proposed Pentagon fiscal 2022 budget failed to provide shipbuilding funding to start such a fleet expansion. The new shipbuilding plan sought to retain many of the traditional types, while building a significant new unmanned fleet, referred to in the report as "uncrewed platforms"—underwater, surface, and air. The navy proposed a total battle force in 2045, including unmanned or uncrewed vessels and logistical ships, of between 398 and 512 platforms, 77 to 140 uncrewed platforms, and 321 to 372 traditionally manned ships.

Of even greater immediate importance for those in Newport News, the proposed fiscal 2022 budget supported eleven aircraft carriers, with about $2.5 billion for carrier work, including, of course, the *Kennedy*. The requested funding financed the fifth "installment" of detailed design and construction for the *Enterprise* and the fourth for the fourth Ford-class carrier, the *Doris Miller*.

The request also included about $166 million dollars to address "unique technologies" for the Ford-class carriers, including research and development for the integrated Digital Shipbuilding (iDS) transformation "to upgrade the digital data environment." The navy said it would save additional money in fiscal 2022 through the reallocation of funding for more efficient ship construction of the *Kennedy*.

When the *Kennedy*'s systems are all turned over to the navy, engineers will be able to monitor them digitally with wall-size screens busy with displays, thanks to all the automation and sensors on *Kennedy*. By then, of course, most of the cables, pipes, valves, and other equipment will be hidden by bulkheads or insulation and pretty much inaccessible.

As the Kennedy CHENG reminded her engineers, "Everything right now is not insulated. You walk and follow a piece of pipe all the way through." That's something you couldn't get from sitting in a chair and looking at a screen. When you're eyeballing those pipes and wires, you can also scan pressures and temperatures, checking to make sure there are gauges in the right places, especially in the air-conditioning plants. Just as important, everyone needed to understand the pre-light-on steps, that is the procedures for turning everything on correctly. Shipbuilders would show sailors exactly how to do that.

Imagine, again, a city constructed from scratch and then the utility crews go back and street by street, block by block, turn on the lights, open the water lines, connect the cabling, and check out the other systems folks need to get through their days and nights. At this moment, shipbuilders would discover whether all their sweat, blood, and steel had paid off, whether they really had a warship or some overpriced metal.

Every carrier comes together differently, and a Ford-class ship like the *Kennedy* featured unique elements as part of the design, but even so, CHENG Owens could count on lessons learned from CVN 78 *Ford* to help plan the *Kennedy*'s transition. CVN 79 sailors had started to use the *Ford* as a training carrier, a floating schoolroom of sorts, with all the hands-on experience that, well, they could get their hands on. Chief Aviation Boatswain's Mate (Equipment) Luis Linares, from Petén, Guatemala, the catapults' leading chief petty officer for *Kennedy*, got himself a crash course in EMALS, which was very different from the cat systems on the Nimitz class he had worked on throughout his career—"more complex," as he told other sailors.

Kennedy plank owner Cheyenne Scarbrough, a reactor electrical technical assistant, saw firsthand the benefit of putting CVN 79 sailors on actual watch teams and real ship drills. That experience would better prepare the crew and the Kennedy for operations.

The navy recruited thirty-nine of CHENG Owens's engineers to provide some extra eyes for damage control for CVN 78 as the ship embarked on one of the most dangerous and potentially destructive operations that a navy ship could conduct in a noncombat scenario. The *Kennedy* damage-control sailors received training unique for a crew aboard a newly built carrier—and a lot more than they bargained for.

The *Ford* headed south with about three thousand sailors and civilians aboard, under the command of Captain Paul Lanzilotta, who took over from Captain J. J. "Yank" Cummings in February 2021. A graduate of the Massachusetts Institute of Technology, he commissioned through the Naval ROTC program and had served in some way, shape, or form on a number of carriers, including *Stennis, Eisenhower, Washington, Reagan, Truman*, and yes, the first *Kennedy*. Like Cherry Marzano, Lanzilotta had also served as the commanding officer of an amphibious ship, the USS *Arlington*.

From the bridge, he watched the ship's steady southern progress through the Atlantic Ocean as the *Ford* made its way to a spot off the coast of Jacksonville, Florida, carefully chosen to make sure the carrier operations disrupted no marine life, particularly migrating mammals.

During the trip, the vessel's damage-control teams trained incessantly on the same drills they had practiced now for about two years, wearing blast suites, firefighting gear, and other special attire. Every carrier crew, of course, trained and retrained for damage control. But the drills now included some special training for a once-in-a-class operation. The last time a carrier faced such tests was during the Cold War.

Out in the Atlantic, Lanzilotta started to carefully monitor the sea, weather, and overall general conditions at four every morning—he had four hours to make the go or no-go decision. Days passed. When the conditions achieved the perfection he needed—cloudless skies and calm seas—the Ford slowed to about twenty knots, about the speed of a car through a busy neighborhood, to keep a rendezvous with Trials and Research Vessel NAWC 38, a baby of a ship compared to the carrier, about 150 feet long and a bit over 130 feet wide. Built about four decades ago, the ship served mostly as an offshore supply vessel. It towed more destructive cargo at the moment for the sole benefit of the Ford. Not a cloud obscured any part of the horizon. The supply ship chugged gently on the gentle swells, none of them rising above three feet, as required for the scheduled dangerous operation. The mandates ensured safety for the ships, the crews, and marine mammals. From the bridge of the Ford, NAWC 38 looked like a large toy, its tether stretching into the depths behind the little ship like a taut tail—a tail with a mighty nasty sting. At the end of the tether, the supply ship towed a bomb of sorts, a forty-thousand-pound explosive charge—roughly equal to a 3.9 magnitude earthquake—to be detonated near the Ford to see if the ship could take a punch. Starting June 18 and ending August 8, the Ford and its crew survived three underwater blasts, each successively closer to the carrier.

Navy and shipyard officials fully expected the Ford to survive with no serious damage or operational stoppage, but in truth, no one knew for sure what would happen when the sea bombs went off so close to the ship's hull.

Explaining the shock testing later, Lanzilotta explained, "Water is pretty darn incompressible." After that detonation, the explosive wave would slam the ship like a high-speed underwater train. Power outages, flooding, fires—the crew prepared for any and all of that. While shipbuilder and navy engineers simulated the blast impact on all the new automated systems baked into the *Ford*, they truly possessed no absolute insight into what havoc the explosion might wreak. The last carrier subjected to the shock trials, in 1987, the *Roosevelt*, operated on analog systems. Computer-controlled systems, like those on the *Ford*, tended to react badly to sudden impacts. On the other hand, the navy anticipated collecting reams of data from all those sensors and systems, providing they didn't all just go dark. The navy also had rigged the carrier with additional cabling, cameras, and other sensors to capture even more information.

Sailors and steelworkers aboard the *Ford* would never forget that first—or second or even third—blast. Over the all-ship public address system, called 1 Main Circuit, or 1MC for short, the strong, clear voice of Oregonian Lieutenant Abbie Ortman, the officer of the deck and the "countdown coordinator for shock," counted down to zero.

With each descending number, stomachs knotted. Fists clenched. Lips pursed. Most crouched and bent at their knees as trained. At the moment of the explosion, the ship rocked as though perched on a giant jackhammer. Darkness enshrouded Lieutenant Ortman and others on the bridge as a column of frothy water rose from the sea toward the top of the carrier tower, past its bridge wing. No one there at that time expected to be plunged into darkness. It lasted only one heart-thumping moment—and then the sun again lit their consoles— momentarily dead with the shipwide loss of power from the blast. Within seconds, though, backup systems kicked in as the displays flashed alarms to check for casualties and other concerns.

Deep in the ship, it sounded like the inside of a barrel when someone set off a small stick of dynamite, and it felt like driving over a rutted, potholed backwoods road at a hundred miles an hour on rubberless steel rims. Debris, insulation, and anything not fastened down tight overhead fell on those below.

Everyone appreciated their training—and earplugs. Commander Tabitha Edwards, in charge of damage control central, watched her engineering panel flicker out for a second, and then the displays, straight out of a gamer's fantasy, snapped back to life after barely an instant. Monitors, sensors, and systems all responded as designed— better, in fact. The ship's automated and sensored-up systems made it much easier and quicker to take stock of any damage.

The only way to truly catalog any damage, though, was to physically inspect the ship, and teams of sailors, including some from the *Kennedy*, started to make their way through the vessel, wearing blast suits and helmet-mounted flashlights. The blasts knocked out shipboard communications, so they carried walkie-talkies. Making their way through the shadows and intermittent glow from battle lanterns, emergency lighting, and other any other illumination available, the sailors assessed the situation. Forty-two groups of sailors—calling themselves the Fighting 42—fanned out over the expansive city-ship to check for cracked welds, smoky billows, and anything they thought needed further inspection. After weeks of training, they knew their way around the ship in the dark. No one panicked—although the smoke caused some anxiety. It turned out to be clouds of debris and dust. No fires. The damage-control team had expected to be working for several hours to get the *Ford* operational. It took only about one hour. The ship conducted helicopter flight operations on the afternoon of the final shock test.

A bit more than a week after the final test, the *Ford* pulled back into Newport News Shipbuilding for what the navy called a planned incremental availability, or PIA—six months of tweaking, fixing, and modernizing to ready the carrier for its first real overseas deployment, planned for 2022. One shipyard supervisor in particular watched the *Ford* dock at Pier 2 with keen interest—Derek Murphy. While no kin to the Mathews Murphys nor a Tidewater master shipbuilder in the classic meaning of the term, Murphy still boasted a naval and shipbuilding pedigree worth noting. His interest in the *Ford* mooring at Pier 2 rested in his current job at the yard: the manager of the CVN 79 program, the supervisor in charge of *Kennedy*, now that the yard entrusted Mike Butler with shepherding CVN 80 *Enterprise*.

Murphy took over the program in February 2021, knowing he faced a real challenge in getting the waterfront to accept him. Not only did workers see him as a come-here, but of the worst possible type—one with shipbuilding ideas from other programs, from another part of the country. He moved to Newport News from Pascagoula, Mississippi, where he worked at the Ingalls Shipbuilding shipyard as the program manager for the construction of US Coast Guard National Security Cutters.

"How do you say no when you can go from the smallest ship program in the company to the biggest in the blink of an eye?" Murphy semi-joked after arriving in Tidewater.

On arrival, Murphy needed to introduce himself and win over the James River shipbuilders. Being from a company cousin provided little collateral with steelworkers from 8888. "Ingalls" did not necessarily ring a sour note, but it turned off some. Murphy wanted to avoid resting on those laurels, being the guy who came up from Ingalls to run a carrier program. Instead, he focused on his other career, which included a great deal of carrier experience.

He'd spent decades flying jets off carriers in A-6 Intruder bombers and EA-6B Prowler electronic attack jets—predecessors to Super Hornets and Growlers. In Newport News, Murphy talked up his navy days. "I flew off carriers, and I want to thank you for your hard work," he told workers—words sure to sound like music to their ears. "You build great ships," he continued. "I never built a carrier, but I did call them home for an awful long period of my life."

In the yard, of course, waterfront workers show respect for

whomever is in charge. "Mister Murphy, would you look at this" quickly replaced the "Hey, Mister Butler" calls that echoed previously throughout the yard.

But Murphy treated the supervisors, the master shipbuilders with decades of experience, with a bit more deference, relying on his previous military training for just such a situation. Ensigns, for example, never told a roomful of experienced chief petty officers how things needed to be, not if those young officers wanted to become older officers.

Above all, he needed to show humility. He called *Kennedy* program leaders into a room and told them, "I'm not here to tell you how to build carriers. You're the best on the planet. We've all got a job to do, and I'm just part of it. I've got my part. You've got your part."

He made the same kind of points when he first left the navy and started working at Ingalls years before, jumping into the cutter work without a lick of shipbuilding experience, after years of flying jets off carriers or helping Northrop Grumman design other aircraft for the ships.

After starting the job and properly introducing himself to the waterfront, Murphy set out to introduce himself to the carrier *John F. Kennedy*. His first day, he spent three hours on just one deck. He could get through most of a cutter in that kind of time. Walking around a carrier under construction, being the one responsible for it, required a completely different mind-set than living on a carrier or flying off one. The only connection he enjoyed previously with the reactor and propulsion spaces on a carrier when he flew jets off them was to send some gullible new pilot down there—in full flight gear—for the nonexistent flight simulator. Reactor crews always got a laugh out of that one.

When Murphy walked through a carrier while flying jets off them, he never walked through every compartment with a flashlight, all the time, à la Don Doverspike.

He jumped into the new job with both feet, literally—doing morning stretches with the waterfront workers before making his morning rounds on the ship, walking into compartments to talk to the crews there, take pictures, and make notes on the work being done. With no chance of seeing the entire ship in even a day, Murphy picked out a zone, a system, or a compartment to focus on for a shift. He dedicated a whole day, for example, to inspecting the island.

Sometimes, as he prepared to start his rounds, he'd hear, "Hey Mister Murphy, can you look at this?"

"Yeah, let's go," he'd call back. So much for his daily rounds that day, but so what? He wanted the workers to see him as being as approachable as possible, particularly with the return to facemasks due to the Delta variant spikes. Murphy saw all the fogged safety glasses on his workers wearing the masks and agreed that, for some shipbuilding jobs, mask wearing ranked right up there with being waterboarded. A conservative by nature and a strong believer in making your own choices about things like health care, he still supported getting a vaccine. He and his family got the shots and, joking with Cherry Marzano one day, Murphy noted that both aviators received plenty of mandatory vaccines throughout their naval careers. The new *Kennedy* manager assured workers about getting vaccinated, telling them, "The company is not going to allow anything to happen to any of you—you are the company."

Fighting through the Covid spikes, Murphy and his crews prepared the *Kennedy* for the ship's next milestones. The yard expected to start up the propulsion plant, but security prevented any public notice of when the nuclear reactors were to come online. The major milestone anticipated during the 2021 fall involved the turning over of 25 percent of compartments at the end of September, giving the navy ownership of a quarter of the ship.

Navy crews already stood watches on the ship during the late summer of 2021. "If you went down belowdecks, into the finished spaces," Murphy told folks, "you'd ask, 'Why can't we go to sea?'" But the higher up in the ship you went, the more work needed to be done, and work was now starting under the single-phase delivery plan. Murphy now anticipated a third-quarter 2024 delivery.

Meeting *Kennedy* milestones became even more achievable, in Murphy's mind, when the *Ford* pulled into the next pier. With so much done already on CVN 79, it was hard to tell it apart from CVN 78. Viewed pierside, the two ships, side by side, looked like carbon-copy carriers, ready for action. True, *Kennedy* needed a lot more work before even starting up its power plant, but the sight of two Ford-class ships moored to piers in the water inflated the chest of more than one steelworker. On a practical note, waterfront workers and sailors moved much more easily between the ships, like a transfusion of people, experience, and lessons learned. Excited to get his first glimpse of a completed Ford-class ship, Murphy walked aboard CVN 78.

Besides 78 and 79, also in the yard at that time, in various stages of carrier completeness, the waterfront in August 2021 tended to CVNs 65, 73, 74, 80, and 81.

As Murphy put it, "You can hold 81 in the palm of your hand right now."

Back in Washington, DC, in the Navy Yard, Rear Admiral Downey chalked up the *Ford*'s phenomenal shock-trials recovery to the computer modeling and design, hardening the digital-centric ship for such assaults. Those systems, now proven on the *Ford*, would be improved on the *Kennedy*—and right down the line.

Once again, with the transition to the Biden administration, there remained some question about the length and strength of that lineage. A major determining factor would be naval leadership, still tarnished from the chaos of the Trump years. As the navy formed and rolled out its budget, it had no secretary to protect its turf—an absolute necessity in the shark pool of Pentagon spending. Biden took even longer to name a new navy chief than he did to propose a budget, much to the chagrin of everyone concerned about the service. In the HASC, Republican Congressman Gallagher went so far as to publicly recommend Democratic Congresswoman Luria for the job—just to get someone named.

Finally, soon after the budget's release, Biden nominated Carlos Del Toro for the post, and he sailed through Senate approval to be sworn in as the seventy-eighth navy secretary on August 9, the day after the *Ford* finished its final shock trial.

Born in Havana, Cuba, Del Toro immigrated to the US in 1962 as a refugee with his family and settled in New York, where he grew up in Hell's Kitchen and graduated from public schools before earning a bachelor of science degree in electrical engineering from the Naval Academy in 1983. Commissioned as a surface warfare officer, he served twenty-two years in the navy, including a stint as the first commanding officer of the guided-missile destroyer USS *Bulkeley* (DDG 84), as well as the engineering department head aboard a destroyer and an aircraft carrier during Operations Desert Storm, Desert Shield, and Provide Comfort. He also had served as special assistant to the director and deputy director of the Office of Management and Budget in the Pentagon, where he helped manage the budgets of the Defense Department, State Department, Central Intelligence Agency, Defense Intelligence Agency, National Reconnaissance Office, and Peace Corps. He retired from the navy as a commander and, in 2004, founded SBG Technology Solutions, serving as the company's CEO and president as it did work on programs in shipbuilding, AI, cybersecurity, acquisitions, space systems, health, and training.

Telling sailors and marines in an online post he had a bias for action, Del Toro named the navy's most pressing challenges as the four Cs—China, Culture, Climate, and Covid. In taking the new job, Del Toro made it clear in public messages that the navy needed more to do its job.

As our nation shifts from a land-based strategy over the past twenty years fighting the wars in the Middle East to a more dominant maritime strategy in the Pacific, particularly in our efforts to deter China, I do believe that our Navy and Marine Corps team will need additional resources to be able to fully field the combat effectiveness we will need as a nation. China continues to develop sophisticated military capabilities to include surface, air, and undersea platforms while demonstrating aggressive behavior that flouts the rules-based order, threatening regional stability and security. I believe that it's imperative to make the right investments and modernization. We can't continue fighting the wars of yesterday; we have to work toward fighting the new wars of tomorrow.

But where did carriers fit in that mind-set—as war tools of yesterday, or tomorrow's tools as well? Del Toro told senators during his nomination hearing, "The strategic environment is rapidly changing due to the pace and fielding of technologically advanced missiles and other weapons, such as cyber and space, designed to reduce the United States Navy's advantages at sea. As such, I believe that the navy and marine corps team should critically look at all alternative platforms, to include alternative aircraft carriers."

Well, an alternative aircraft carrier was still an aircraft carrier. Del Toro also defended the *Ford* to the Senate: "While the Ford class has faced challenges with development and construction delays, the program incorporates advances in technology such as a new reactor plant, propulsion system, electric plant, Electromagnetic Aircraft Launch System, Advanced Arresting Gear, machinery control, and integrated warfare systems that are expected to increase lethality, and lower life cycle costs through reductions in maintenance and manning requirements. . . . I believe the Ford class carrier will prove to be a critical combat enabler."

But the doubts about the carrier's relevance, effectiveness, and overall return on investment lingered on around the Beltway. As the US conducted its public and controversial withdrawal from Afghanistan in August, just a few months after the Biden administration released its budget and Congressionally dissatisfying long-term shipbuilding plan, Congressman Adam Smith, the Wisconsin Democrat who ruled over defense funding as the powerful HASC chairman, called into question the carrier's future all over again, even resurrecting the Mark Esper navy plans, believed to be dead and buried.

During a public online appearance for the DC-based Brookings Institute think tank, Smith, at first, seemed to be lauding carriers, touting "the ability of the carrier to [provide] presence in the Persian Gulf or South China Sea [and] to send a signal about the commitment to that area. Presence still matters."

But his next compliment came with a bit of a backhand: "I think there is some utility in the carrier, even if you cannot get as close," because, he explained, "the changing pace of the technology, especially missile technology . . . puts our ships in a [vulnerable] position."

As a result, he said, "The aircraft carrier isn't going to be able to get as close as it used to be able to. You can move unmanned systems off it. You can envision it to move platforms closer, if not as close as you used to."

But then, he pointed out, US lawmakers must consider the biggest issue about buying carriers. "Are there other ways to get close to the fight or get unmanned systems closer without that cost of twelve-billion dollars?"

That analysis is still being done, he noted. "I haven't done the math on that personally."

Shipbuilders and naval carrier program officers knew what to expect—another carrier study, to match the dozens done to date. In some ways, carrier supporters welcomed another study, believing the ship shone when compared with other options. But Smith also indicated that the new studies, any new analysis regarding shipbuilding plans, promised to be gamed differently this time around, focusing, as Esper did, on overall capability, instead of specific ship counts and numbers.

"For most of my time in Congress, we've been obsessed with numbers," Smith lamented. "If somehow we were able to write down on a piece of paper we plan to have a four-hundred-ship navy—as long as we talk about numbers of ships, it's all good."

Esper, instead, wanted to find first the needed capability of the overall force, and then develop the fleet necessary for that requirement. "That has a lot of resonance going forward," Smith predicted. "That vision is incredibly relevant right now."

That vision, carrier supporters recalled, included a slowdown in carrier production and a price increase in overall construction, which threatened to make it even more difficult to justify the carrier expense with so many doubters about the ships' attributes in future fights.

Of course, shipbuilders and navy carrier supporters boasted loud and often about the capability—the potential—for *Ford*, *Kennedy*, and other Ford-class ships to shine in future conflicts, because of the ships' advanced technology. The ships needed to prove that as they headed out to sea.

As Del Toro pointed out during his nomination hearing, "The reliability growth of key systems will increase as those systems continue to mature and operate during at-sea periods."

He said he would examine "the navy's strategy to improve system reliability growth for key systems and review how the navy and industry are addressing lessons learned to ensure they are being applied to the fullest extent to increase reliability and drive down costs of follow-on ships."

And, of course, CVN 79 *John F. Kennedy* was the first follow-on ship. Each morning, before his daily rounds, each and every shift, Derek Murphy tried to drive home to his waterfront workers the importance of their work. He talked to them by the side of the pier, in ship spaces, anywhere it made sense.

The steelworkers, though, knew of their worth. Deemed "essential workers" during the pandemic, they used that designation to demand better contract provisions in the winter of 2021—even threatening a strike to drive home their point.

But Murphy focused more on making the best carrier, not the most money. "This ship will go into harm's way," he reminded them. "Whether it's in World War III with China or off some coast chasing pirates, nobody knows. Remember this—the last commanding officer of the *JFK* hasn't even been born yet, that's how long this ship will be in service."

He used some body English to emphasize his message. "You guys need to take pride in what you're building and what you're doing. If you're not enjoying what you're doing, step back and take pride in what you're building."

Then, he gave them a homework assignment. "When you walk off the pier at the end of the shift, turn around and look at this carrier. You're building the most powerful warship in the world. What you do matters—to your family, your community, the navy, and this nation.

"China has been trying to build one of these and it can't. The Soviets tried and could not. Russia is still trying and it can't. No one can do what you do—they just don't know what you know."

ABBREVIATIONS GLOSSARY

1MC—1 Main Circuit, shipwide public address system

A1B (reactor)—aircraft carrier platform, first core design, Bechtel

A4W (reactor)—aircraft carrier platform, fourth core design, Westinghouse

AAG—advanced arresting gear

AFL-CIO—American Federation of Labor and Congress of Industrial Organizations

AWE—advanced weapons elevator

CAG—Crisis Action Group

CDC—Centers for Disease Control and Prevention

CHENG—chief engineer

CNO—chief of naval operations

CO—commanding officer

CVN—carrier vessel, nuclear

CVNX—carrier vessel, nuclear, unnamed new class

DBR—dual-band radar

DDG—destroyer designated, guided missile

EASR—Enterprise Air Surveillance Radar

EEOC—Equal Employment Opportunity Commission

EMALS —electromagnetic aircraft launch system

GAO—Government Accountability Office

HASC—House Armed Services Committee

IAM—International Association of Machinists

ICAN—Integrated Communications and Advanced Networks

iDS—Integrated Digital Shipbuilding

LCS—littoral combat ship

LL—liberal leave

LPD—landing platform dock

NAACP—National Association for the Advancement of Colored People

NLRB—National Labor Relations Board

NR—Naval Reactors

OSHA—Occupational Safety and Health Administration

PCU—precommission unit

PPE—personal protective equipment

PSA—Peninsula Shipbuilders Association

PSA—post-shakedown availability

SASC—Senate Armed Services Committee

SSBN—submersible ship, ballistic missile, nuclear

SSN—submersible ship, nuclear

STD—short-term disability

USWA—United Steelworkers of America

X—Newport News Shipbuilding trades designation

XO—executive officer

ACKNOWLEDGMENTS

First, I must acknowledge the steelworkers of Local 8888 in Newport News, Virginia, as well as everyone else at Newport News Shipbuilding. They opened their doors and their lives to my probing eyes, ears, and questions. In particular, I want to thank Bill Bowser for revealing so much about himself and those turbulent times so I could understand. I hope he finally found peace in his passing.

Without the help, generosity, and patience of Mike Butler, the "shipyard mayor" of the construction of aircraft carrier *John F. Kennedy*, this book would never have been possible.

I also want to especially acknowledge those in the US Navy connected with the *Kennedy* construction, especially Captain Todd "Cherry" Marzano, the "navy mayor" of the *Kennedy*, who kept me on the right course.

I owe a special debt to my agent, mentor, and friend—James D. Hornfischer—who pushed me to write the book he knew I could. He may be gone, but I still feel his presence at my elbow every time I sit at a keyboard.

I want to express my eternal gratitude to Mary Dempsey, Troy Graham, and Jerry Millevoi for their sharp eyes, keen sense of judgment, and brutal but necessary critiques.

I also remain indebted to Mauro DiPreta, who recognized the potential for this work, exhibited amazing flexibility when things went sideways, and provided the insightful editing needed to whip these pages into shape.

Others who deserve a special shout-out include the following: Edward "Big Ed" Elliott (RIP), Edward "Little Ed" Elliott, Lee Murphy, Jason Murphy, Mike and Dianne Davis, Mike Petters, Arnold Outlaw, Dan Sunderland, Wayne Deberry, Sam Carper, Odis Wesby, Don Doverspike, Charlie Holloway, Warren and Sheila Outlaw,

Jennifer Boykin, Lucas Hicks, Aaron McCann, Jordan Patterson, Matt Phoebus, Charles Spivey, Erica Brinson, Derek Murphy, Duane Bourne, Beci Brenton, Bob Haner, Captain Danny Hernandez, Rear Admiral Michael Manazir, Admiral Buddy Yates (RIP), Commander Mike Prudhomme, Machinist Mate Nuclear First Class Aaron Zevenbergen, Nuclear Electrician's Mate Chief Petty Officer Kevin Stambaugh, Ensign Cheyenne Scarbrough, Legalman Chief Rasha Shankle, Master-at-Arms First Class Chief Petty Officer Kristi Dennis, Boatswain's Mate First Class David Kuefler, Information Systems Technician First Class Chandler Ragland, Damage Controlman First Class Caleb Peterson, Lieutenant Commander Valerie Greenaway (the *Kennedy* training officer), Chief Dominique Sherrod, Machinist Mate First Class Petty Officer Brandon Nowlan, Machinist Mate First Class Petty Officer Evan Raycraft, Rear Admiral James Downey, Captain Joe Klopfer, *Kennedy* Command Master Chief Thaddeus Wright, Senior Chief Jayme Pastoric, Captain John "Yank" Cummings, Captain Brett "Chopper" Crozier, Rear Admiral John Meier, Captain Clayton Doss, Commander Angela Owens, Captain Paul Lanzilotta, Rear Admiral Charles Brown, Rear Admiral Brian Antonio, James "Hondo" Geurts, Dave Hendrickson, Tina McCloud, Bryan Clark, Brent Sadler, Randy Forbes, Robert Holzer, Joseph P. Kennedy III, those at the Newport News Mariners' Museum and the Peninsula Rescue Mission, Cody Spencer, and Chris Mathews.